data
to life

MAR 2014

Data to Life

MATT REED

JOSS LANGFORD

Coelition

Version 1.0

Published by Coelition, London

Copyright © Coelition, 2013

ISBN 978-0-9576094-0-2

Contents

Preface

It is now hard to imagine what life was like before digital technology became ubiquitous. The internet, email, text messaging, the World Wide Web, social media, mobile smartphones and e-commerce have all brought a revolution in our daily lives. These technologies have arrived in successive waves, and the next wave is now imminent, as multiple technologies mature and converge. Wearable devices that are always connected, 'big data'-driven predictive analytics and the Internet of Things are poised to create numerous personalised services in health, wellness, commerce and leisure.

Such services have the potential to deliver huge benefits to individuals, communities, businesses and society at large, but there are real and growing concerns about what the technologies will be doing in the background. The sheer volume of data that personalised services will collect from an individual sounds uncomfortably like the world described by George Orwell in his dystopian novel *1984*. Many believe that if we are not careful big data could lead to Big Brother.

Debate on these issues tends to polarise between those who see data-collecting enterprises as inherently sinister and those who argue that privacy is already dead so anything in pursuit of profit is fair game. We believe there is a middle way, one that facilitates the enormous benefits these new technologies can deliver without creating an unacceptable level of intrusion.

This middle way is an international standard for collecting and handling personal data that will provide both privacy for consumers and opportunities for enterprise. Coelition is the organisation founded to develop and publish this standard. It is a not-for-profit consortium which, in its pre-competitive phase, will craft the first edition of the standard. At that point, the consortium will become fully open, inviting any interested party to become a member and help direct its future progress.

This book describes the solution we propose and in so doing shows how we can bring data to life. It provides impetus for the early stages of the consortium and a public-domain forum to discuss the ideas on which it is built. We hope the book will be relevant to anyone with an interest in how we collect and process information about our daily lives, not just to those who go on to use the Coelition standard. Equally, it should be completely possible for a developer to implement the standard without having read this text.

Over the course of three years we have built a software engine that has helped us develop the insights contained in this book. This is the original engine on which the Coelition standard was modelled and it will become a blueprint for future versions. Our work established both the deep rationale behind the concept of an 'atom of behaviour' and the practical importance of the idea in real-world situations.

Throughout the project we have found certain principles indispensable in guiding our day-to-day decisions. We have brought these together at the end of the book in the belief that they provide a strong foundation to any work of this nature. *Data to Life* is both a description of the journey we have made in creating Coelition and a call to action for those who want to unleash the benefits of personalised services while keeping the privacy of individuals paramount.

Matt Reed & Joss Langford
2 September 2013

Acknowledgements

This book would never have arrived in your hands without the guiding pen of our editor Jill Hopper. Thank you, Jill, for your professionalism and support throughout.

Our story is largely built from the monthly meetings of the team who crafted the original software engine behind Coelition – the Gang of Four. We have a debt to Paul Bruton, Idris Eckley, Brian Newby and Ruediger Zillmer who challenged and constructed these ideas with us.

Thank you also to Alannah Warner for help with the concepts about time and to Ben Childs for the illustrations in that chapter.

Finally, we are grateful to our families – especially to Kate Langford for her diligent reading of multiple drafts, and to Dawn, Lorna and Ben Reed for their ongoing support.

Daily Life 1

THIS is a story of daily life. It is a story about how we think about daily life, how we measure it and how we can learn from those measurements. It is about the daily lives of real individuals, told with unprecedented detail and accuracy. Lastly, it is a story of how scientists, engineers and business people can come together to explore a whole new frontier of research and understanding.

In the last half-century, the boundaries of human endeavour have been pushed ever outwards. In 1969 men first walked on the moon, in 2012 robots collected rock samples from the surface of Mars and now Voyager 1 is on the verge of leaving our solar system. Yet at the same time another, quieter adventure is unfolding, and a different type of boundary is being breached. Increasing numbers of us are waking up to the idea that, by turning our attention inwards rather than outwards, and decoding the minutiae of our personal lives – how we brush our teeth, how we travel to work – we could crack some of the most pressing problems of our time. Climate change; spiralling deaths from non-communicable diseases such as cancer, diabetes and heart disease; the collapse of public health and transport systems – the answers to these seemingly intractable issues may be, quite literally, under our noses.

The way we live the details of our daily lives is fundamental in defining who we are. It is not what we say about ourselves, in conversations or on Facebook, that captures our true nature. These are carefully edited versions of ourselves as we would like others to see us. It is what we actually *do* – routine by routine, task by task and habit by habit – that is the most reliable guide to who we really are. And yet most of us are blind to the meaning of these daily patterns. We believe we do one thing, yet in fact we do another. Most of our solitary actions are accompanied by thought processes that wander from one fleeting topic to the next, while our social interactions are filled with subconscious elements from our animal past. There is a profound disconnection between what we say or think we do, and what we actually do.

Imagine an alien coming to Earth and seeing human beings for the first time. It is the actions we perform and the behaviours we exhibit that our visitor would immediately register, without needing to understand anything of the underlying intentions. Our alien friend would no doubt categorise us not by our physical appearance but by our habits and movements. Some human beings would rise before the sun and others work through the night; some would travel at regular times to the same places every day, while others would mostly stay at home; some would be found in large groups and others mainly alone; some would regularly smoke plant matter and others would not. To the alien eye, these patterns of daily life would swiftly separate us into numerous different types and even allow further differentiation of individuals within those types. They would reveal not only our schedules but also our characters and, ultimately, our life expectancy.

Technology is already providing us with the ability to observe ourselves with the equivalent of this 'alien eye'. In the mature economies of the West, we have an array of tools to capture and interpret the details and patterns of daily life, and the large consumer markets of China, India and South America will soon arrive at a similar point. Two trends – our ability to record data about our lives, and the integration into our social structures of technologies that need this data to operate – are irreversible. Mobile phones, personal computers, satellite navigation systems and credit cards all require knowledge of our daily actions and identities to provide us with the services we demand. As data scientist Sandy Pentland says: 'We live our lives in digital networks. We wake up in the morning, check our email, make a quick phone call, commute to work, buy lunch. Many of these transactions leave digital breadcrumbs – tiny records of our daily experiences.'

With the increasing sophistication of sensors, data algorithms, data networks and cloud storage we will inexorably create more information that will dig ever deeper into our personal lives. Many people see this as a huge opportunity to improve our individual quality of life and learn what makes us tick. The Quantified Self movement, for example, encompasses a wide range of individual enthusiasts who record their own personal behaviours over time, to identify connections that elude their normal powers of perception and reach a better understanding of themselves.

Big data

We write this book not just from an academic standpoint but from a position of personal experience. For example, one of the co-authors, Joss Langford, experimented with a wrist-worn sensor for a number of years, and this book draws on his findings. The sensor Joss wore is similar to a watch, but rather than telling the wearer the time, it records their physical actions and environment in precise increments of one-hundredth of a second. Although it can measure time more accurately than any individual could, it understands nothing of the minute-to-minute lived experience of its user. The challenge of reconciling these very different truths is central to our mission.

Unfortunately, the wrist-worn sensor creates another challenge when seeking to understand daily life: a data challenge. Each day that you wear the device on your wrist you could create more than 100 megabytes of data. By the end of a week this is more digital data than could be held on a typical personal computer in 1995. By the end of a year you will have amassed more data than all humans had created together from the start of civilization to the beginning of the information age. This amount of material comes under the statistical definition of 'big data' – very difficult to handle with traditional approaches and needing specific, dedicated tools to extract useful information from the sea of bits and bytes. In fact the only tools equipped to handle this volume of data are those originally created by scientists for looking at the stars, trying to predict the weather or decoding human DNA.

Alongside this technical definition of big data there is another more emotive and society-driven definition, which Sandy Pentland has described. Under this alternative definition, big data is the trail of fragments of information about ourselves that we leave, just like footprints or breadcrumbs, during our daily interactions in a digital world. Pentland says: 'Big data is increasingly about real behaviour, and by analysing this sort of data, scientists can tell…whether you are the sort of person who will pay back loans…if you're likely to get diabetes.'

And so big data could easily beget Big Brother. We have lurched into a fully connected world and are seizing the huge benefits it will bring, but are currently unwilling or unable to grapple with the downsides. The problem cannot be ignored any longer. As individuals and societies it is imperative we make choices about how to manage the dangers. The current giants of the internet succeeded in the Wild West of the initial digital boom: they made their own rules when there were no rules. Advanced data protection and personal privacy protocols should not be seen as a roadblock to these corporations' progress, or that of others. On the contrary, developing rigorous standards will allow us to secure and capitalise on the benefits in a balanced and sustainable way. This book describes how we have met the challenge of big data head-on, negotiating a route through various sets of proposed privacy rules to a place that has the best interests of individuals and of wider society at its core.

You may see the focus on digitised personal data as bizarre, self-obsessed or unnecessary. However, two in three adults in the US will track some aspect of their health and most of us will maintain a diary of some description. Even if we have no interest in gathering such information ourselves, the technology we use will be doing so automatically, and we need to think about who has access to this information, how it is stored and what it can be used for. Although we may wish to refuse to share anything about our lives, the reality is that our behaviour affects others in profound ways. Our personal health and the health of our planet depend on tiny, individual daily actions. A single cigarette doesn't kill and one tank of fuel in a vehicle will not end the world. People kill themselves and harm the planet one action and one day at a time. Reversing the effects of these practices must also be tackled one action and one day at a time.

Daily Life

Brands and personalisation

In our own lives we can all think of individuals who have influenced the way we see the world and our role within it, prompting us to make sweeping changes. However, achieving this effect across a mass of individuals requires agency of a different kind. In a consumer society, it is to trusted brands that people turn when they need help in making choices. Brands are often positioned as artificial constructs of large corporations but in fact it is consumers en masse who build brands by repeatedly choosing them over other offerings. We need strong, trusted brands to help make our lives easier, their promises becoming shorthand for a collective understanding of a complex bundle of functional features and emotional reassurance. Large corporations may legally own a logo and try to influence how it is perceived, but it is consumers who truly control its meaning.

So, with their ability to proliferate overnight and influence the actions of millions, brands are the ideal agent for mass behaviour change. Writing for Unilever, one of the world's largest branded goods companies, in 2011, even environmental campaigner Jonathon Porritt grudgingly accepted the role of the brand in sustainability: 'A bit of me still recoils in a mixture of repugnance and disbelief at the idea that it's going to be the world's leading brands that will rescue us from today's slow but inexorable slide into ecological disaster.'

Porritt was contributing to a short publication on behaviour change that showed how the owners of global brands want to create a better dialogue with their consumers. These brands have realised they need to reconnect directly with consumers, without intermediate retailers, if they are to remain relevant in the future. To build viable relationships with consumers they need to get close to them as individuals rather than viewing them as arbitrary demographic clusters.

A holistic approach to consumer engagement is the inevitable outcome of an increasingly personalised digital marketplace. The digitisation of commercial and social transactions is dissolving the boundary between the physical and virtual worlds and the digital data trail of choices we leave behind us has already enabled a degree of online customisation. This can be seen in the emergence of search-engine personalisation, tailored e-commerce and social network tie-ins. The logical next step is to use personalisation to start a conversation.

It is essential for any digital behaviour change system that it 'talks' in an accessible way – that its outputs make immediate, intuitive sense to the user and are rooted in the context of their day. This is a factor acknowledged by both behaviour change scientists and Quantified Self practitioners. Once these insights have been delivered they can be used as a trigger for positive change, if only we can arrive at a common standard that allows individuals, communities and businesses to work together. We propose such a standard at the end of the book. It is not merely a theoretical proposal – our understanding has been forged by building real systems of hardware, servers, networks, software, visualisations and people, following the principles of the standard. We know it works.

Data to life

This book describes our journey through everyday life and how the digital systems we all rely on interact with us day by day. It challenges our thinking about how we reduce our complex daily patterns to data, how we treat that data and then how we bring the data to bear on our lives in a fruitful way. It is a journey from life to data and from data back to life.

In the next chapter we start our story with an exploration of the smallest elements that make up our daily lives. We then consider how we can organise knowledge to build intelligent systems capable of creating meaningful connections to people. The issues of privacy and data protection are examined alongside the power that shared data can offer in changing how we live our lives. Finally, we propose tools and standards that will enable us to access this power safely and transparently.

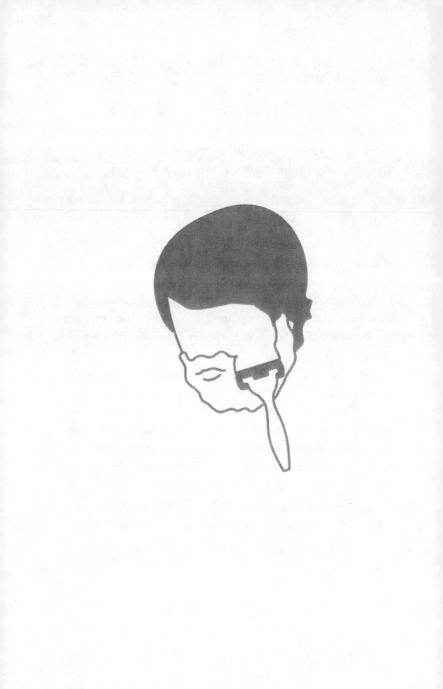

Atoms 2

'Let us record the atoms as they fall upon the mind in the order in which they fall, let us trace the pattern, however disconnected and incoherent in appearance, which each sight or incident scores upon the consciousness. Let us not take it for granted that life exists more fully in what is commonly thought big than in what is commonly thought small.'

So wrote Virginia Woolf in 1925, in her essay *Modern Fiction*, which analysed the attempts of contemporary novelists to 'come closer to life' by considering the minutiae of everyday existence. Woolf herself, and her contemporaries James Joyce and Gertrude Stein, spent their writing lives attempting to break with traditional narrative forms and pin down the connections between individual consciousness and great upheavals in society and politics. This focus on an honest accounting for the 'atoms' of everyday life led James Joyce to write *Ulysses*, a huge novel of a quarter of a million words describing in minute detail a single day in the life of his protagonist Leopold Bloom.

Woolf's words have a renewed resonance for us today. How could we make real her desire to 'record the atoms' of everyday life? Apart from the practical problem of how we might be able to capture this type of atom, there is a deeper question: what, if anything, *is* an atom of life?

In the physical sciences, nearly four hundred years of scientific research have established that the whole of the universe is made of an uncountable number of tiny entities, called atoms. Although there are an almost limitless number of discrete atoms, they only come in a small number of different flavours or types, each with a set of well-defined physical characteristics and chemical properties. This is known as the 'atomic hypothesis' of physical science – the universe is granular and all things are made of atoms.

Nothing is more fundamental to the whole of chemistry and biology than the atomic hypothesis. In 1964, the American Nobel Prize-winning physicist Richard Feynman highlighted how crucial this hypothesis is for modern science:

'If, in some cataclysm, all of scientific knowledge were to be destroyed, and only one sentence passed on to the next generations of creatures, what statement would contain the most information in the fewest words? I believe it is the *atomic hypothesis* ... that *all things are made of atoms –*

little particles that move around in perpetual motion, attracting each other when they are a little distance apart, but repelling upon being squeezed into one another. In that one sentence, you will see, there is an *enormous* amount of information about the world, if just a little imagination and thinking are applied.'

Although the term 'hypothesis' may suggest that this idea is conditional or unproven, for almost all practising life scientists and chemists the atomic hypothesis is treated as if it were a solid and uncontested fact. Still, the idea remains counter-intuitive. Matter does not readily reveal its basic granularity. What we actually see and feel around us is a continuous world.

An atom is almost too tiny to imagine. Typically, atoms are about one hundred trillionth of a metre in diameter and, without specialist scientific equipment, are well beyond our powers of perception. However, each of these atoms also has an internal structure that determines its chemical and physical properties. These internal structures are not infinitely variable: there are only so many ways they can be put together to make an atom. This regularity leads to the small number of naturally occurring chemical elements. Fewer than one hundred chemical elements, such as oxygen, carbon, gold, silicon and hydrogen, occur naturally on Earth.

The ramifications of this are mind-boggling. All the diverse forms of life on Earth, and the matter in the rest of the universe, are constructed by combinations of these elements. Fish, trees, human beings, stars, gases, water, snow, apples and oranges are made from this one set of ingredients, combined in a dazzling number of ways.

Inspired by the atomic hypothesis of physical science and Virginia Woolf's literary and psychological insights, we propose an atomic hypothesis to account for the events of everyday life. Although an uncountably large number of events happen in the lives of billions of people around the world each day, only a very limited number of elemental behavioural atoms make up everyday life. What makes our individual lives unique is not a huge range of types of behavioural atom, but the infinitely diverse ways we string these atoms together into the rituals and habits of our days. Our quality of life is not determined primarily by exceptional, one-off events such as marriage or births, but instead by the fine-textured fabric of everyday life. Like Virginia Woolf, we do not 'take it for granted that life exists more fully in what is commonly thought big than in what is commonly thought small'.

Before moving to a more detailed exploration of this atomic hypothesis, we illustrate the basic idea with a very simple example. Figure 2.1 defines a crude and non-exhaustive selection of some of the typical atoms of behaviour that may be involved in a person's everyday life.

This incomplete list reveals some important features. The behavioural atoms need to be big enough for people to identify them easily as discrete and distinctive events.

Atoms

Event	Atom
A	Wake up
B	Make a cup of tea
C	Wash my face
D	Pay for purchases
E	Work on a computer
F	Phone my daughter
G	Go to bed

Figure 2.1 A crude and non-exhaustive selection of some of the typical behavioural atoms found in a person's everyday life.

If they are too small then they are not memorable enough. For example, the atom 'Make a cup of tea' is readily identifiable. Even in a regular tea drinker's average day, there will be no more than ten or so of these atoms. If we tried to find smaller elements than this, they would be too small to be remembered – each stir of the spoon involved in making a cup of tea is too fine-grained to be of use. The atoms should be elemental in nature – that is, there are only a limited number. They should be relevant to the lives of ordinary people. Some of them may be short in time duration but very noticeable ('Pay for purchases') and others may be of longer time duration but harder to notice or remember ('Work on a computer').

Each of the behavioural atoms listed in Figure 2.1 has an internal structure that we could delineate if we wanted to create a detailed recipe for the atom. In fact, physical atoms also have a complex internal structure, but for everyone except atomic physicists this internal structure is of little interest and is essentially hidden when atoms interact. Similarly, for most purposes the internal structure of a behavioural atom – how many stirs of the spoon were involved in making a cup of tea – is irrelevant. In contrast, the structures created by the mutual interactions between different types of atom can be fascinating.

For example, suppose we know that Bill typically has the following atoms in his day:

A, G, B, E

This is of limited use. Presumably he wakes up (A) before he makes a cup of tea (B) but unless we can put these atoms into the order in which Bill usually does them we are in the dark.

Once we are able to put these atoms *in time order* we can get a better, though still nowhere near complete, picture of Bill's day:

A, B, E, G

If we now try to describe the interactions between these atoms by putting some intervals between them, we obtain a more useful picture of Bill's activities. What's more, if we can put a time-marker on each, we can pick up the beginnings of a pattern in Bill's life:

A   (06:45)
B   (06:50)
E   (08:35-17:00)
B   (15:30)
B   (19:30)
G   (23:15)

A complete daily ritual does not readily reveal its basic granularity. The concept of a behavioural atom began as a thought experiment during the early stages of our work. It throws up an immediate welter of questions and possibilities, and it merits a detailed explanation. Most of the remainder of this book describes what we have learnt whilst taking this thought experiment and making it work.

One of the first things we wrestled with was the question of how many atoms would be required to understand a life. Our initial impulse was to assume that our existences are sufficiently varied that there must be many thousands. But we soon came to reconsider. We know that our lives are not *infinitely* varied – and most are dominated by some form of routine during which we do certain things over and over again. No matter how sophisticated we think we are, we all spend the bulk of our time each day on tiny, ordinary activities: sleeping, eating, drinking coffee, shopping, calling friends, driving to work, texting our partner, washing our hair, using Facebook, cooking a meal...

This is exactly what James Joyce and Virginia Woolf discerned a century ago. Human existence, by and large, is not formed of the transcendent moments poets speak of, but of repetitions and patterns. Using a rough order-of-magnitude estimate we found that as few as ten atoms could already delineate some interesting structures in an individual's day. One hundred atoms were enough to describe a month of an individual's life. One thousand atoms could capture virtually everything an individual would do throughout their entire life, or describe up to a month for many different people. With ten thousand atoms we can delineate the lives of anyone. Having a total of between one thousand and ten thousand atoms is also practical, as this is a manageable quantity that can be read by a human being as well as a computer.

From now on we will use the term atoms to refer to atoms of behaviour. So when you read something like 'all of Bill's atoms' please suppress the physicist in you that is trying to visualise the oxygen, hydrogen and carbon that make up his cells.

Having done some groundwork, we can turn to the characterisation of the atom, as set out in Figure 2.2. This characterisation follows the classic journalist's questions of what, when, how, why, who and where. These questions will also form the next six chapter headings of this book.

Field	Notes	Example
What	What type of event was this?	Make a cup of tea
When	When did this event begin and what was the duration?	07:50 GMT 120 seconds
How	How was the event recorded?	A specialised Android app
Why	Why was this event recorded or which event preceded it?	Bill was prompted to record it
Who	Who performed this event?	Bill Smith
Where	Where did this event happen?	Bill's kitchen

Figure 2.2   The internal structure of an atom of behaviour.

Atoms

What 3

THE urge to ask: 'What is that?' is a deeply human one. It develops naturally as we grow up, and fuels the drive we all have to name and classify objects and concepts in the world around us. In this chapter, we apply this urge to the task of identifying and classifying the atoms of everyday life.

Creating a comprehensive list of every atom that could exist in the life of any person on the planet is a daunting challenge. How do we spread the net wide enough to capture the diversity of lifestyles we see across the globe? How do we make the list of atoms large enough to be comprehensive, but small enough that we end up with a manageable number? How do we make the listing comprehensible to both humans and computers?

Like many things in life, there is no more elegant way of proceeding than by simply trying our best and making a start. Anyone can do it, beginning at any point in the day, and continuing until no more atoms come to mind. The interested reader may want to try this themselves now: write down as exhaustively as possible all the actions you perform during a typical day (get up, wash face, go for a run, have a cup of tea, etc.), ignoring when and where you do them. The items you note will inevitably be influenced by your own life history, your family circumstances and your broader cultural context – including religious, national and other factors – but nevertheless it is a pretty straightforward job.

However, working through this exercise immediately raises some interrelated issues:

– The list we have created has no internal logic or structure other than the order in which we have recorded the atoms. This simple ordering does not help us understand the whole list. If atom 31 is no more related to atom 32 than it is to atom 59, then we cannot easily navigate the list.

– We soon find we have either direct duplicates of atoms or, more likely, atoms that are so similar to others already on the list that we feel uncomfortable about keeping them separate. We find ourselves wanting to merge two very similar atoms into a single entity that covers both. This tendency to cluster things that have similar characteristics is a natural part of how humans deal with the multiplicity and complexity of the world. We are pattern-forming creatures.

– On reflection, some of the atoms we have included may strike us as being of a different type or kind from the others. For example, at first it does not seem unreasonable to include 'feel sad' as an atom. But when we compare that with 'have a cup of tea', we may become uneasy and conclude that these are different *types* of thing.

What

- Even after spending some time on this very basic collection, it is probable that we will still end up with an incomplete list of all the atoms that could occur. How do we decide when we are finished, or even when we are ninety or ninety-nine per cent there?

These issues are really important. It is all very well for us to claim that everyone has a circumscribed number of atoms in their life, but unless the full list exists, in a usable format, the idea has no impact. In the rest of this chapter we describe how we have tackled this basic challenge and describe the resulting structure: the Classification of Everyday Living (COEL). It must be noted that, although we refer to this system as a 'classification', technically it is in fact a taxonomy, which we discuss further below, since it both names and orders the atoms.

In building a taxonomy, it is fundamental that all items within it are of a similar type. In philosophical terms, they must be of the same *category* to allow true comparisons to be made between them. Many existing taxonomies of everyday life describe the world in terms of objects. However, the nature of an atom of behaviour is that it occurs at a defined time – and to assign a time to an object is, in philosophical terms, a category error. The book that my grandfather gave me on my eighth birthday remains the same object. Although it provides a sense of nostalgia for me when I hold it today, it is an object of the here and now with a history, rather than something that existed only in the historical past.

Based on a number of years' practical experience, we have found that for the COEL to be applicable and useful it must be a classification purely of *events*. In it we define an event as: 'a transient and time-bound activity that can be objectively recorded by a person or a device'.

It is crucial to note that although the COEL refers to the everyday lives of humans, its remit specifically excludes thoughts, feelings, intentions and emotions. No doubt for virtually all the atoms listed there are associated thoughts and intentions, either conscious or unconscious, but none of these could be objectively recorded by a person or device. They are therefore, in our definition, *not* events.

Suppose that we have an event – a tree falls. This type of event is definitely transient, it has a specific time when it happened and, if we were close to where the tree fell, then we could objectively record the event happening. Associated with this event we can imagine that there may be a whole range of non-events.

Tree falls	=	the event
Bill decides to cut down the tree	=	the intention leading to the event
Jane is sad to see the tree falling	=	an emotion following the event

These three statements are very different. Simply recording the event 'Tree falls' gives no insight into either the
intention that led to the event or any emotions induced
by it. Although both intentions and emotions are of real
human interest, they do not fit into our taxonomy. It
would be a category error to include them.

If an event happens routinely in our lives then we
may feel that we want to record a non-event. For example,
I might habitually have a midday meal, which I happen
to call lunch. Although the repeated appearance of this
event in my life is of some interest, the missing lunch on
a particular day may be more so, and we are naturally
tempted to record the absence. However, we cannot
include such a non-event in the COEL. We can only
detect a non-event when we look at a person's life over a
longer timescale – it is the non-appearance of a lunch
event after twenty-eight days of midday meals that is
notable and perhaps important.

Non-events, emotions and intentions can all be
recorded by initiating a simple interaction with the person
concerned. When we make an utterance or a recordable
statement, an observable event has occurred which can
then be coded as an atom. So, for example, an individual
can be prompted to answer a question such as 'How do
you feel this morning?' using a scale of happiness or
selecting an option from a list. We could then code the
response as an emotion, while also preserving a reference
to the original input in order to understand the context.

Data to Life

Classification, naming and taxonomy

Before we can quantify something, either by counting or measuring, we first need to be able to discriminate it from other types of thing: we need to *classify*. We also need to be able to name it: we need a *nomenclature*. A scheme that includes a consistent and interlinked classification and nomenclature is known as a *taxonomy*.

One of the earliest, and still most widely used, of all taxonomies was devised by the Swedish botanist Carl Linnaeus in 1753. Linnaeus wanted to find a way of simplifying the long and complex Latin names then used for plants into 'trivial names'. He used a simple two-part name – much like the combination of a given name and family name that many of us use (such as John Smith). This idea of a binomial nomenclature was so powerful that almost immediately, and ever since, plant and animal species were given a two-part name (such as *Taraxacum officinale* for dandelion).

The first part of a binomial name identifies the *genus* to which the species belongs (genus being related to the term 'generic'); the second part identifies the *species* within the genus (species being related to the term 'specific'). Linnaeus's binomial nomenclature helps avoid confusion. If we use common names to refer to a species, we may find that in different parts of the world these refer to completely different entities. Take the common name 'field mouse', for example. In Europe, a field mouse is one of several species in the genus *Apodemus*, in North

America it is a small vole in the genus *Microtus*, while in South America it will probably refer to a small mouse in the genus *Akodon*. Other common taxonomies include international aviation safety standards, grocery store layouts, the Dewey Decimal System for books and the Periodic Table of Chemical Elements.

If Linnaeus's taxonomy offered biologists merely a means for naming individual species, it would have limited utility. The brilliance of his system was that it gave rise to a hierarchical classification scheme: it helped biologists cluster groups of separate species together, on the basis of shared characteristics, to create the higher-level structure of the *genus*. Several of these can be further combined to form a *family*, another rung up on the ladder. The full hierarchy of biological species has eight levels that run, from top to bottom: *domain, kingdom, phylum, class, order, family, genus* and *species.* This means that in principle any of the 1.2 million currently known species, or any one of the estimated 7 million species yet to be named, can be correctly allocated a slot in the hierarchy by asking seven good questions and getting seven good answers. The structure of the hierarchy also implies that it would be much more unusual to find a completely new *phylum* or *class* of organisms than to identify a new species.

The Classification of Everyday Living (COEL)

The COEL is a new event-based taxonomy for classifying and naming the atoms of everyday life. It was constructed according to a set of design principles, as follows:

1 Individual atoms sit at the bottom of a logically clustered hierarchy. Atoms that have certain similarities are kept together.
2 Category errors are to be avoided. The list includes only events, where an event is defined as 'a transient, time-bound activity that can be objectively recorded by a person or device'.
3 Each atom in the list is distinct from all other atoms, that is, they are mutually exclusive.
4 Together, the atoms should completely cover the whole of everyday human activity, that is, they are collectively exhaustive.

Building a comprehensive event-based taxonomy from scratch is ambitious and time-consuming, requiring subject-matter expertise and a holistic view of the problem. In particular, ensuring the list is both mutually exclusive and collectively exhaustive (that is, complete and without overlaps) is very demanding. A classification that meets both of these demands is known as a 'MECE' structure, and creating one requires the use of multiple iterations of divergent and convergent problem solving. After several years' work we have arrived at a point where we are satisfied that the COEL is indeed a MECE taxonomy of everyday events.

The most logical way to describe the COEL structure is from the top down. Before we begin the description, however, the reader needs to bear in mind that the fine-grained and most interesting detail is at the bottom of the hierarchy, at the level of the individual atoms.

At the top level of the COEL tree we have defined around thirty *clusters* of event classes that go together. The name of each cluster has been chosen to be intuitive for users of the classification, and each has a specific definition fully describing its contents. Some of these clusters inevitably have a much richer structure than others, since certain elements of daily life contain more variation than others. For example, the Personal Care, Eat and Pastimes clusters all have complex lower-level structures, as these are the main ways that people define themselves and how they spend their time. The Work cluster has a much simpler lower structure.

Below the level of the clusters come three further levels: *class*, *sub-class* and *element*. Each of these, in turn, has a maximum of nine further levels below it. This structure is shown schematically in Figures 3.1 and 3.2.

Every imaginable atom of everyday life is assigned a unique name and numerical code, making the overall structure readable by both humans and computers. The COEL is also as culturally agnostic as possible, with the potential for universal application across different human cultures and times, provided that a culturally 'translated'

Classification of Everyday Living (Key)

Clusters

30

Classes

O

30 x 9 = 270

Sub-classes

△

30 x 9 x 9 = 2,430

Elements

◇

30 x 9 x 9 x 9 = 21,870

Figure 3.1   The hierarchy of the Classification of Everyday Living. Reading downwards, each layer adds detail: clusters, classes, sub-classes and, finally, elements. In total the hierarchy can contain nearly 22,000 individual elements.

Classification of Everyday Living (Structure)

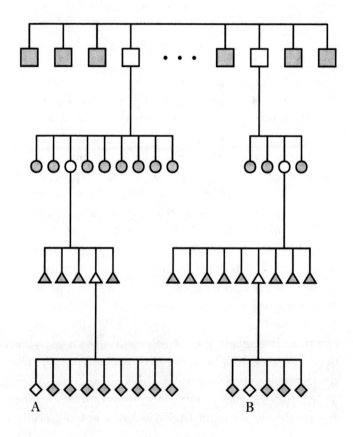

Figure 3.2   Two elements, A and B, can be identified precisely by reading the listing of sub-class and class upwards. They can be simply differentiated by reference to the cluster. For example, A could be a 'Food' event and B a 'Work' event.

version is developed, with appropriate descriptions of the common events and a set of specific extensions.

Visualising atoms

The COEL represents a unique approach to obtaining data inputs that can be used to capture the daily life events of ordinary people. One of the natural corollaries of creating ways to collect input information has been to consider how the COEL can best deliver output information – that is, what is the most helpful way to give people feedback about their behaviour? What we have learnt is interesting. Most people do not want to see information about their daily lives displayed as if it were a stock market graph in a newspaper. While line graphs, and pie charts may be standard in spreadsheet programs and business reports, we wanted a far warmer and more human way to communicate the insights collected by the COEL. And so we used it as the basis for a set of emotionally engaging icons representing everyday life events, commissioned after studying the Isotype approach developed by Otto Neurath and his co-workers in the 1930s. Because these icons are intimately linked to the rigorous structure of the COEL, they are not just 'pretty pictures' but a coherent lexicon of instantly recognisable symbols (see Figure 3.3).

The current version of the COEL represents a significant asset for those who are interested in recording, analysing and interpreting quotidian events. By sharing this asset with a broad range of partners via the Coelition standard, as described in Chapter 10, we believe we can create a uniquely rich resource for all those interested in everyday life.

We believe the Classification of Everyday Living can become as widely used as the biological taxonomy first created by Linnaeus.

Data to Life

Figure 3.3 *A subset of icons illustrating atoms of behaviour contained in the Classification of Everyday Living.*

When 4

OUR subjective experience of time passing is a deep mystery, essentially because 'events are perceivable but time is not'. For technical researchers in the field of neuroscience this is a knotty problem, explored with MRI scanners and brain-imaging experiments. For most of the rest of us it is a vast and uncharted territory of unknown shape: we don't even know what we don't know.

Part of the problem is that we commonly use just one word, *time*, to describe both *physical* and *personal* time, which are in reality two quite different things. Clocks measure physical time as an objective quantity, with no reference to human experience. Over the past four hundred years, scientists and engineers have developed increasingly sophisticated clocks for measuring elapsed time periods; today's ordinary quartz wristwatch is more accurate than any instrument in existence before 1900.

Since the 1960s the world has used Coordinated Universal Time (UTC) as the standard measure of physical time. It is based on an atomic clock where 'the second is the duration of 9,192,631,770 periods of the radiation corresponding to the transition between the two hyperfine levels of the ground state of the caesium 133 atom'. Computers generally use a system known as Unix time, based on the number of seconds that have elapsed since midnight on 1 January 1970 as measured by UTC.

However, no matter how accurate our wristwatches or atomic clocks, they do not help us make reliable judgements about time as it is actually experienced. Our time-reckoning abilities are influenced by learning, culture and personality. What is more, the passing of personal time fluctuates according to our state of mind, mood, environment and even body temperature.

We have all experienced brief periods when we are oblivious to or acutely aware of time passing. We say that time flies when we are having fun, but drags when we are bored. After a particularly intense bout of work, emotional stress or enjoyment, we feel we have packed a wealth of experience into just a few days. Yet at other times the weeks seem to slip by unnoticed. 'Where did the spring go?' 'Did we really have a sunny summer or are my memories mixed up with previous summers?' 'Look how much my son has grown this year.' It is common to feel that time slips by even faster as we get older. Personal time is as slippery and elastic as physical time is inert and rigid.

If physical time is cold in comparison to the richness of personal time, it is the world around us that weaves them together – albeit in a non-linear and often confusing fashion. Our personal time is metered and modified by our perception of the physical world as we interact with it and in it. Our subconscious learns from an early age the inexorable, destructive power of raw physics – the glass breaks but will never reassemble, and the spinning top always slows. These entropic effects provide us with lessons about the one-way direction of time: we know all too well that we cannot see the future or alter the past.

However, life's ability to rebuild and renew itself provides us with another, more hopeful arrow of time. Although the physical conditions of the seasons, such as altering light and temperature levels, are dictated by the planet's tilted movement around the sun, without life on Earth, spring and autumn would be indistinguishable, with similar weather and length of days. The surges of vernal growth and autumnal decay, powerful markers of time that provoke corresponding changes in our mood and behaviour, are provided by the trees and animals around us. And although the residual heat of the day offers a material difference between sunrise and sunset, it is life that provides the music and scents that give each its identity.

If we accept, then, that personal time plays a fundamental role in assessing the world, we are left with the challenge that computers cannot deal with it; for the moment, physical time is the only yardstick a computer can use to interact with the complex, lived experience of human beings. When employing a digital system we must find viable ways to translate between the world of fluid personal time and that of rigid, consistent physical time.

Data to Life

Time-stamping atoms

To record an atom of behaviour we first need to place it within the succession of other atoms and determine its duration. We can then go on to pinpoint it in physical time. In some cases we can use measurement approaches that live entirely within the realm of digital, physical time. For example, information about when a phone conversation started and the length of the call is best reported by the phone rather than by the caller.

For many behaviours, even the most meaningful ones, individuals need help from digital systems to recall and translate events from their personal time to physical time. This is because our reliability in remembering and recording activities varies, as shown in the list below:

1. Events are normally in the right order.
2. Time-reckoning accuracy will vary for different activities.
3. Regular daily activities are easily remembered.
4. Regular non-daily activities are often forgotten on a specific day (Did I fill the car with fuel on Tuesday or Wednesday?).
5. Vague activities have vague times and durations (I watched television for a few hours).
6. Anchoring events have more accurate time stamps (I collect the kids from school every weekday at 15.10).
7. Co-ordinated time with people or infrastructure has more accurate time stamps (I met a friend at 12.30 for lunch; I caught the train at 11.32).

When reconciling a personal diary account with physical time, beginning with the more definite atoms within a day helps us infer the duration of the vaguer activities.

This task of moving from life to data is undeniably difficult. However, moving in the opposite direction – turning an array of atoms back into a meaningful narrative – is equally daunting. It took novelist James Joyce nearly a decade to describe a single day in the life of his protagonist Leopold Bloom, as recounted in his masterpiece *Ulysses*. Joyce tackles head-on the interplay of physical and personal time by minutely exploring the rhythms of Bloom's inner life amid the march of external events.

Daily cycles

The anthropologist Claude Lévi-Strauss used the concepts of 'diachronic' and 'synchronic' time to distinguish between those aspects of life that can be studied with a historical perspective of time (diachronic) and those that benefited from a perspective *across* time (synchronic). Much of our everyday behaviour is time-based but is best understood outside of linear time. Our patterns, rituals and habits are often cyclical in nature and we can best interpret them through the concept of a typical day rather than any specific 24-hour period. Figure 4.1 shows a scheme for visualising a typical day. The use of a continuous spiral with a 24-hour circumference allows us to look

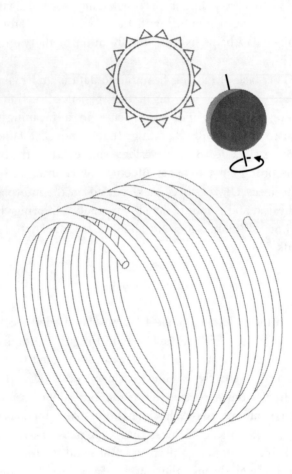

Figure 4.1   Above: The 24-hour rotation of the Earth with respect to the sun. Below: Wrapping linear time into a spiral, one loop for each day over seven days, allows us to look 'through' time at a typical day.

Figure 4.2   Near-body wrist temperature shown on a 24-hour cycle averaged over 28 days. Sleep period as determined by activity is shown in darker grey.

'through' time and provides a first step towards a more readable representation.

The sun is with us for an average of twelve hours per day, whereas adults sleep for approximately eight hours. There is, of course, a wide range of sleep patterns that would be considered normal. We all know whether we are larks or owls, and have a sense that we used to be more owl-like as teenagers. Despite these differences, we all start and end our days with the same atoms of behaviour: waking up and going to sleep. If we were to compare the activities of two individuals on a certain day, we might want to know about their sleep patterns but to contrast their waking hours we would need a common framework to which they would both relate.

A wrist-worn sensor which records movement, light and temperature can help determine aspects of the sleep cycle – lack of motion can signify sleep and light levels may indicate the 'setting' of our body clocks. The fluctuation of body temperature over a 24-hour cycle is particularly useful, as can be seen in Figure 4.2, which shows data collected by Joss. The human body lowers its core temperature at the onset of sleep by increasing the temperature of the extremities to shed heat. On waking this process is reversed.

If we imagine the circle in Figure 4.2 slotting into the spiral in Figure 4.1, we can see that it has been rotated so that the mid-sleep time is at the base. This sleep period is referred to as the chronotype, the technical term indicating the time of day when an individual prefers to

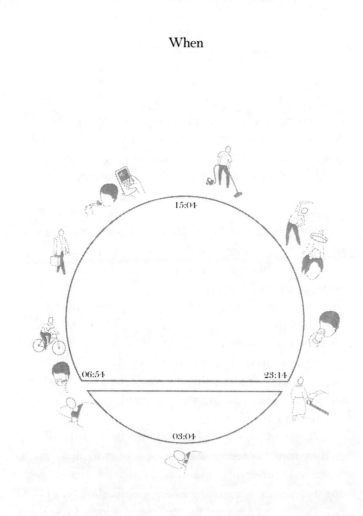

Figure 4.3 A representation of a 24-hour day with waking and sleeping hours separate. Icons illustrate activities within the day.

sleep (the lark/owl phenomenon). As well as using a wrist-worn sensor, an individual's chronotype can be determined using sleep diaries, free-living measurement, questionnaires and even genetic analysis. (It is worth noting, however, that a person's chronotype and the number of hours they spend sleeping are largely un-related.)

We are awake and conscious for two-thirds of the day and for the remaining third we shift into another world – a discontinuity we are all deeply aware of. This disjunction can be useful in helping us visualise personal time. In Figure 4.3 the waking and sleeping hours have been separated to create an asymmetric form. It is important to note that the top point of the arc does not represent midday, i.e. noon, but the middle of the waking day – twelve hours after the chronotype. This point is where we are least alert during the day and coincides roughly with the traditional time of siesta.

This two-segment circle allows us to make direct comparisons between the daily lives of individuals with different chronotypes, and to be consistent across differ-ent time zones while using a single Unix time reference. All we need to know is the individual's local time zone and chronotype.

The scale of the waking and sleeping arcs will vary between individuals, depending on how much time they spend sleeping. The scale does not need to be linear, since its primary task is to communicate the succession and duration of atoms of behaviour. You can begin to see how this simple heuristic chart, mapping a few details of an individual's life, functions well as a bridge between a computer's physical time and our personal time.

The general approach of looking 'through' time can be further developed to portray seasonal cycles. Figure 4.4 takes the daily spiral and wraps 365 of them into a year, as the Earth completes a full rotation of the sun. The motion of the moon, important in the breeding cycles of many life forms, is not shown on this representation but the lunar phases would create a candy-stripe around the annual solar spiral.

Longer-term variations of behaviour and factors that influence activity can be depicted on this yearly basis. Figure 4.5 shows actual light exposure (made up of both intensity and daily duration) captured by Joss's wrist-worn sensor. Light received at the wrist is affected by many things such as weather, levels of indoor and outdoor activity, and clothing. However, the annual light levels experienced by someone living at a latitude of 51°N can be clearly discerned in this chart.

Profane time

This chapter has dealt at some length with personal and physical time. However, there is a third category of time that we need to consider before we move on. This is 'profane' time: the social measure that we use to organise our everyday lives. It has no cosmological basis but holds very deep cultural meaning, both in its labels and

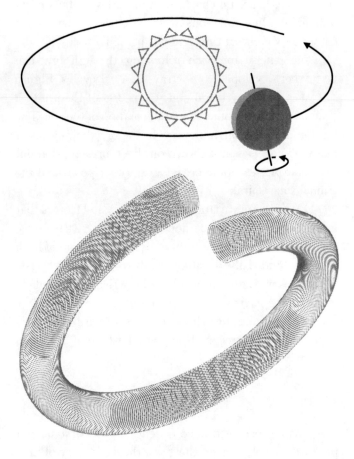

Figure 4.4   Above: The annual and daily rotation of the Earth around the sun. Below: 365 rotations of the Earth complete one rotation around the sun to create this double spiral, allowing us to look 'through' time at a year.

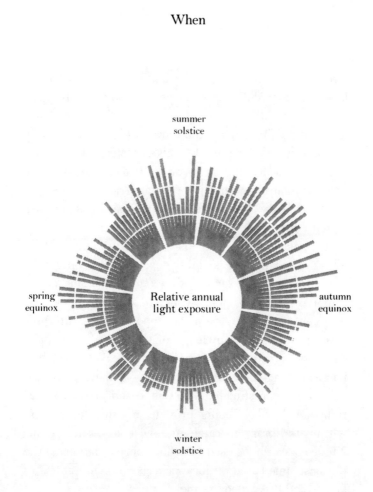

Figure 4.5 Light exposure at the wrist for a whole year. Months divided equally into ten epochs with averaged values .

structures. At one level, Sunday and Monday are arbitrary labels for consecutive 24-hour periods; however, they are loaded with different significance for different societies. The seven daily loops in the spiral of Figure 4.1 provide a representation of a week. This is a cultural structure that imposes strong patterns on our lives in the same way that the monthly calendar does in agrarian societies.

Profane time is integral to our language and provides us with a powerful tool to orientate ourselves, allowing us to make social arrangements with minimum effort. This use of language is highly dependent on context for its meaning. For example, the statement 'come round tomorrow before dinner' would be clearly understood between friends, but almost impossible to interpret precisely without knowledge of the geography, social background and local practice of the speaker. The challenge language poses to the analysis of behaviour should never be underestimated. For this reason, the Classification of Everyday Living provides a dedicated section within its library to catalogue and code profane time. These codes can be used to define approximations of the physical time based on local criteria, allowing atoms to be correctly recorded in time.

When

How 5

HAVING worked out how to structure an atom of behaviour and classify the huge range of behaviours that exist in everyday life, we now need to examine the practical issues that will allow the theory to become a workable system.

The first half of the challenge is to elicit information about the real world. There are three basic approaches to building a digital picture of an individual's life:

1   An individual reports his or her own behaviours.
2   An observer records an individual's behaviours.
3   A device directly measures or infers behaviours.

All three methods require a link between the person and the digital system, which will comprise servers, the internet, the cloud, mobile phones and home computers. However, the crucial connection is not to the computer or the cloud, it is to the Coelition system, containing hundreds of years' worth of human knowledge and experience. It is from this that an individual can draw benefit and to this that they can contribute the value of their own everyday experiences.

The second half of the challenge is to provide meaningful and reliable feedback to the individual concerned. It is to trusted brands that we often turn for this conversation, since they provide us with a personification of a service with which we can have a relationship. Our system is designed to allow individuals to choose the brand or service provider they want to interact with, while maintaining a single engine to do the back-room work.

Eliciting information from people

It is unusual for people to approach a questionnaire with enthusiasm or actively choose to repeat it once completed. For two years from 2009, the website Hunch.com offered twenty pairs of image choices online, such as: 'Chunky chips or French fries?'. The average visitor (who received no service or reward) not only answered these twenty questions, but also opted to go on and answer ninety-three more; in all, more than eighty million people participated. The balance of images, insight, speed and playfulness proved irresistible, and yielded a treasure trove of information about the interconnected patterns of personal preference. We should bear this example in mind when designing a questionnaire: they are the least desirable form of research, but if you have to use one, Hunch.com is an excellent model.

In the field of behaviour change, it is often necessary to gather facts about individuals and their daily habits. However, in general, as humans we do not have the capacity to understand and recall the enduring features of our lives while simultaneously focusing on the minute-by-minute detail of our experience. It takes a purposeful effort to bring our complete attention to bear on the present moment. Indeed, this feat is so demanding it requires specific mental training and is the basis of both psychological techniques, such as mindfulness, and older spiritual traditions, such as Buddhism. On the other hand, reflection – the casting of our mind back over much longer periods of experience, even to childhood – means we fail to fully inhabit the present: we are 'lost in thought' or 'daydreaming'. What is more, our recollection of events is often sketchy or unreliable. Diaries and questionnaires can go some way towards overcoming this difficulty, and can in principle capture immediate detail and integrate this over longer time periods, though they do suffer from issues of completeness and accuracy. Digital devices can also be used to prompt the user to input information as they go about their business, thereby flagging up events that might otherwise be overlooked or forgotten.

The other human approach to collecting information about our lives is through observation. In recent years, ethnographic research techniques have shifted from the field of anthropology into the mainstream of consumer research. The process of close observation and recording produces numerous individual reports that are as detailed as they are separate from one another. Digital technologies in observation studies have so far failed to make a significant impact, but the Classification of Everyday Living (COEL) is set to change this. Clearly defined atoms of behaviour allow observed events to be coded in a consistent manner across different studies. The events can then be digitised within the overall system to create quantitative reports for both individual and multiple studies.

Using devices to elicit information

A machine or device can measure behaviour passively or actively. The passive approaches have a low cognitive burden for the user, for example a background application running on a mobile phone can record when a call takes place and its duration. We have created atom-returning applications that can measure many types of phone activities without any input or interaction from the user. Figure 5.1 shows calls and text messages collated by one such application, running on Joss's Android phone for six months. The calls can be seen to track the working day, while most text messages occur at the beginning and end of the day, revealing that for Joss calls are used primarily for work communications while text messages are a more personal format. It is also evident that Joss does not spend hours chatting to his non-work friends on his mobile – other users would have vastly different profiles. We have also created further atom-returning applications that engage the user, prompting them to complete simple tasks, answer questions or make choices.

These are examples of direct measurements, but it is also possible to infer behaviours that we cannot measure directly. For instance, we might want to calculate the amount of time that an individual spends driving. A direct

02.00 04.00 06.00 08.00 10.00 12.00 14.00 16.00 18.00 20.00 22.00 24.00

Time of day

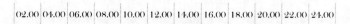 Mobile voice calls   SMS text messages

Figure 5.1 Joss's phone usage, based on six months of atoms and depicted in two-hourly slots throughout a typical day. This shows relative frequency of mobile voice calls and SMS text messages, both sent and received.

approach would be to place an instrument in their car, but if this is not possible we can turn instead to a wrist-worn sensor. This would track the position of the body over time, the motion of the arm and the vibration of the vehicle, allowing us to conclude with reasonable confidence when and for how long the person was driving. Since we want to capture a wide range of distinct atoms, relying on devices to infer certain behaviours is, at times, inevitable. Although some may see this as a constraint, it is in fact a discipline that must be accepted as we move from sensing to measuring.

A sensor that is sampled at a certain rate will create a stream of data points. With computer clock speeds increasing, sensors becoming more widespread, network speeds rising and memory getting cheaper, it is easy to see why we have had an exponential increase of data in society. This big data, however, is just that – data, not information.

Creating information from data is what scientists do. Quantitative methods underpin most scientific research and are generally used in two ways. Firstly, science relies heavily on experimentally or observationally obtained quantitative data to reach valid conclusions about the world. Secondly, it uses data and other information about the world to build quantitative frameworks that can predict, explain and understand complicated phenomena.

Measurement is a defined process, not the willy-nilly gathering of raw data. If it is to be effective, it requires an understanding of the knowledge we are trying to gain, the use of an appropriate sensing technique and a judgement about the number of data points needed to gain an accurate reading. It might be an understandable reaction to take a feverish child's temperature every five minutes but, in reality, no more information is gained than from a half-hourly check.

As discriminators of behaviour, atom-returning devices sit on a continuum from simple to sophisticated. A simple device might be a cigarette lighter that records when it is used, in order to determine patterns of smoking. At the other end of the scale, a person with a laptop and a suitable application could record almost any type of behaviour. In both cases, embedded algorithms are at the core of the behaviour identification process. Even the cigarette lighter runs a simple algorithm determining that two strikes within a few seconds of each other belong to a single cigarette-lighting event.

It will be the development of algorithms, rather than measurement technologies, that will provide much of the progress in the automatic identification of different human behaviours in the coming years. A distinct and important role for algorithms running on general purpose computing devices will be to act as 'interpreters' – taking streams of raw data from existing devices and converting them into atoms.

It is important not to be distracted by the technology within a device when considering functionality. A video camera running a highly sophisticated algorithm to determine walking speed might return the same information as a pedometer. Our atom-based approach is device agnostic, meaning we can future-proof our system for measurement technology and devices that have not yet been invented. We can anticipate products becoming smaller, more intelligent and self-powered but the behavioural atoms and digital interface remain the same.

The range of devices and equipment currently around us with the potential to return atoms of behaviour is surprising:

1 Public infrastructure   (ticket barrier/CCTV)
2 Shared private infrastructure (shower/front door)
3 Fixed computing devices   (PC/digital TV)
4 Portable computers   (laptop/tablet)
5 Phones and pocket devices   (mobile phone/camera)
6 Wearables   (wristband/clip-on)

Wearables

The last category, wearables, is clearly the most appealing when we search for a device that can accompany us constantly without being intrusive or cumbersome. They are closest to our bodies and with us for the longest periods of time during the day. Glasses are comfortable and socially acceptable although, in common with other types of headgear, there are certain social and functional situations where they are removed. Necklaces, bracelets and wristbands are also popular and socially acceptable enough to be worn continuously. Watches in particular are unisex and can support a reasonable bulk without becoming obtrusive. Indeed, taking a long view of the anthropology of wrist-worn articles shows us this may be the best format for digital sensing devices, since they have been a feature of human culture for millennia.

Human beings have been wearing strung beads on their necks and wrists as body adornment for at least one hundred thousand years. Our cultures and even our species have evolved in their presence, imbuing them with deep-rooted meaning. The earliest examples of wrist-worn articles were made from, or strung on, perishable material that has not survived to the present day. The archaeological record of arm bands and bracelets whose purpose can be easily interpreted starts about ten thousand years ago, and their use is widespread, encompassing many different forms, materials and processes. Examples exist from many spheres, including art, society, religion, war, health and wealth. Figure 5.2 gives some examples from the last four thousand years.

It is clear from the high degree of aesthetic finishing of all these examples that the primary purpose of any wrist-worn article is body adornment. Many forms will initially have had specific functions, but the signified meaning is just as important, and can be described using three themes:

1   Protection   (physical and spiritual; non-aggressive)
2   Identity   (personal and expressing individuality)
3   Status   (visible and used for display)

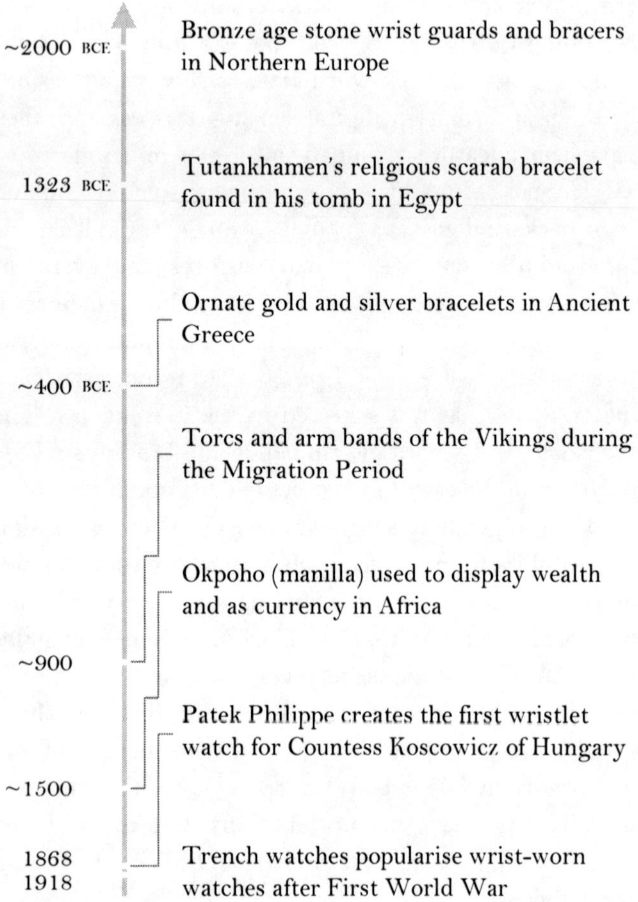

~2000 BCE — Bronze age stone wrist guards and bracers in Northern Europe

1323 BCE — Tutankhamen's religious scarab bracelet found in his tomb in Egypt

Ornate gold and silver bracelets in Ancient Greece

~400 BCE —

Torcs and arm bands of the Vikings during the Migration Period

Okpoho (manilla) used to display wealth and as currency in Africa

~900 —

Patek Philippe creates the first wristlet watch for Countess Koscowicz of Hungary

~1500 —

1868
1918 — Trench watches popularise wrist-worn watches after First World War

Figure 5.2   A timeline with examples to illustrate 4,000 years of wrist-worn body adornment.

How

This anthropological perspective provides confidence that, with care, we can design wrist-worn digital devices that will integrate smoothly into our lives and help us change them for the better.

Closing the loop

Once we have identified the best devices and approaches to elicit information about daily life, we can plan and provide helpful interventions. At the simplest level, merely giving someone objective information about what has actually occurred can provide profound insights and instigate change. It is impossible to gain reliable, subjective information about our quality of sleep, for example, as we are not directly conscious of the passing of the event. Even for events that we are theoretically capable of recording with great accuracy, objective feedback without any emotional baggage can be revelatory.

When we perceive thunder and lightning from an electrical storm with two of our senses, we can calculate the distance of the storm. In the same way, once we collect information from just a few orthogonal sources, it is possible to identify patterns and correlations and create new information. As the number of information streams increases, the possible inferences move quickly beyond the capabilities of our everyday thinking. A system can help us to unravel this into meaningful action.

Figure 5.3 shows the role of the device in recording, measuring or inferring behaviour and sending atoms to a cloud server. The delegation of inference to a layer above the device removes an important constraint in the design of products. It ceases to be necessary for the atom-returning device to be the source of feedback, allowing it to be smaller and require less power. Also, by opening the system up to other information sources, the outputs will be richer and any cloud-based algorithm can respond dynamically to external factors, such as the weather.

Internet of Things

Within the earlier list of possible atom-returning devices, there was no category for everyday appliances such as the fridge. Many believe this will soon change. The ability to create a low-cost connection to the internet provides the potential to network even more of our machines, a world that has come to be known as the Internet of Things. In this world, your fridge and domestic heating system can have machine-to-machine conversations with each other – but what would they say? This question has been explored through Brad the Toaster, created by Simone Rebaudengo. Brad has the ability to tell other machines when he is toasting and has his own Twitter account. Although Brad has been given some cute sayings, all he 'knows' is where he is and whether or not he is being used.

The example of Brad the Toaster highlights the opportunities and limitations of the Internet of Things, showing that unless we know who is using a device we

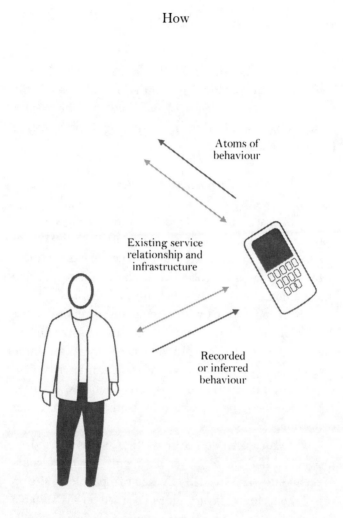

Atoms of
behaviour

Existing service
relationship and
infrastructure

Recorded
or inferred
behaviour

*Figure 5.3   The role of device or machine in the system –
sending atoms to a cloud server and, separately, handling the
feedback path.*

build no understanding in the human world. It also demonstrates that a low-bandwidth connection is suitable for an internet-enabled appliance and that we need a humanising layer, in this case co-opting Twitter with a few pre-canned tweets.

Engineering the system

People interact with a wide range of computing systems, apps and devices that are or could be networked. We can use these apps and devices to determine aspects of human behaviour that are processed in the cloud to provide information, interventions, benefits and services.

Atoms provide us with a template to represent the details of any event. These atoms are constructed in individual packages roughly equivalent in size to a single mouse click. The time stamp of the atom is critical in sequencing information from different sources, while its *how* field alerts the system to its method and means of collection.

In Figure 5.4, we see how the Coelition standard provides a humanising layer between the requirements of technology and the daily lives of real people. It sits on top of the technical layers – it can be used on any suitable device and atoms can be sent through existing channels, such as Bluetooth and SMS – but below the meaning and behaviour change layers, indicating it does not directly relate to a theory or discussion about motivations. It does,

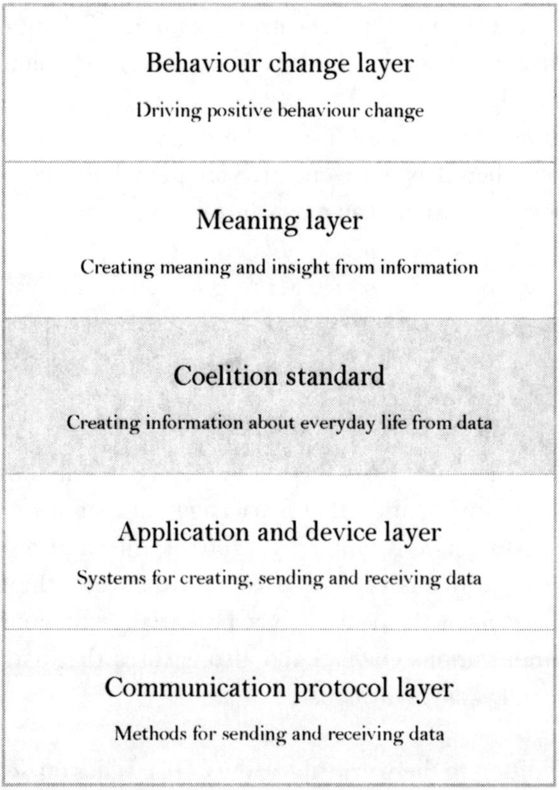

Behaviour change layer

Driving positive behaviour change

Meaning layer

Creating meaning and insight from information

Coelition standard

Creating information about everyday life from data

Application and device layer

Systems for creating, sending and receiving data

Communication protocol layer

Methods for sending and receiving data

Figure 5.4   Coelition standard in context.

however, provide a standardised way to convert those theories and discussions into action.

Embedded in the standard are a number of engineering design principles that ensure the overall system runs efficiently:

- *Distributed intelligence.* If something can be done on a local device, then it is more effective for the whole system to do it there, rather than on a centralised server.
- *Connection between devices and servers should require only low-bandwidth connection.* Making a system that can use the low-bandwidth protocols means it can be used globally.
- *Discrimination should be distributed to the device layer.* Devices should be used to discriminate between different event types.
- *Atoms are generated on devices.* Where possible, devices generate information that is packaged into atoms rather than raw data. Combining control information and payload into this type of 'packet' structure is the most robust means to deal with a potentially intermittent communications channel and also ensures that data can be handled asynchronously.

In addition to these principles, the COEL relies on devices and applications to handle aspects of culture and language. This can be illustrated by the example of food diaries. Whether the food diaries are manual or automatic (new image-analysis approaches can provide food type and quantity information), the atoms are sent as individual food types. So the application could take a culturally specific input of 'chicken burger' and split it into 150g chicken (fried), 50g white bread and 15g mayonnaise, before sending the information.

In this chapter we have seen much of the underlying mechanics required to bring information about the real world into a digital system. The 'how' field of an atom allows us to record the specific means by which information was gained, thus providing the context we need to resolve contradictions and synthesise new information from parallel streams of atoms.

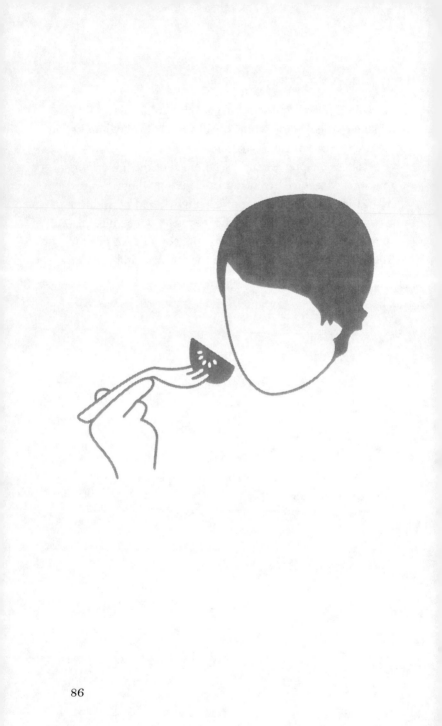

Why 6

OVER the past twenty years there has been a funda-mental change in the availability of information. Ordinary people now have an almost limitless universe of knowledge at their fingertips, aided by sophisticated search engines, high-bandwidth communications and mobile devices that are both ultra-connected and smart.

People are actively looking for experiences, products and services that work for them and their families, and this technological revolution allows them unprecedented choice. Yet, although millions of consumers now have more control over their purchases than ever before, these decisions are just one small element of their wellbeing. Consumption is not enabling them to transform their lives. Most people *know* what they need to do to become healthier and happier, yet do not always act on this knowledge.

There is no easy explanation for this inertia. As we have seen, the Classification of Everyday Living provides a reliable structure for recording *what* a person does. However, it also exposes the difficulty of understanding *why* they behave as they do.

We can try to rationalise our behaviour. It is, after all, a very human tendency to justify our actions and emotions with a common-sense, cause-and-effect expla-nation:

'I was feeling depressed, so I didn't go to the party.'
'He praised me and that made me happy.'

To a greater or lesser extent, we all use this type of rationalisation. Whether it is expressed to others or kept to ourselves, it satisfies our desire for clear and logical links between cause and effect. This belief in everyday causality is often unconscious, but it is an important element in our construction of personal and social narra-tives.

In fact, the task of identifying the main cause of a particular effect is always more difficult than it first appears. In many of the physical sciences, such as engineering, physics and chemistry, cause and effect can often be successfully teased apart using what are known as reductionist methods. Using this approach, complex problems are broken down into a series of smaller component problems, each of which is studied and understood in isolation.

The reductionist approach can be illustrated by thinking about what happens during a game of pool. Imagine a whole game of pool being observed from above. The game is not continuous. It looks like a sequence of individual collisions between the end of a cue and the white cue ball, followed by the moving cue ball causing further collisions with the coloured balls. Over time, some of the balls disappear into the pockets. Although the cue ball keeps reappearing after it has been in a pocket, the coloured balls do not. A reductionist approach allows us to isolate each of these different events and use physics to analyse the dynamics of the ball-to-ball collisions. From these analyses, the general rules governing the collisions can be derived; this detailed knowledge can then be used to model and predict the effects of particular collisions. Here, cause and effect can be understood reliably and in detail, because the collision laws are deterministic – they operate like clockwork. Much of the progress made in the physical sciences over the past four centuries has relied on this type of reductionist approach.

Unfortunately, reductionism is less useful for studying the complex phenomena found in cell biology, physiology, ecology, meteorology, psychology and the social sciences. In these fields, the observed phenomena are not simple chains of cause and effect. Often there are many plausible or possible causes for a single phenomenon.

Complex phenomena are usually composed of entities engaged in a network of two- or multi-way interactions, often with non-linear responses, multiple timescales and positive or negative feedback loops. In these systems, apparently small causes can give rise to unpredictable and surprisingly large effects: this is the origin of the 'law of unintended consequences'.

Trying to isolate just one element of these complex webs is often unsuccessful. Interfering with what looks like an isolated element in order to measure it will often perturb the overall phenomenon. It is therefore often difficult or impossible to use reductionism to unravel how the system as a whole will react to an external change.

Everyday human behaviour is, of course, a complex phenomenon and reductionism cannot help us formulate reliable rules for why people act as they do. Each of us is completely unique. Chance events, life circumstances and external factors all influence our responses, guaranteeing that our individual behaviour is as distinctive as our own set of fingerprints. The list of factors affecting us at any one time is enormously long, and includes both social factors (such as class, economic wealth, personal history, upbringing, and the quality, type and number of personal relationships) and internal factors (personality, belief systems, and states of mind or health).

Most of the humanities and the social and behavioural sciences are occupied with understanding some, or all, of these factors. It is already a difficult task to rationalise why groups of people behave in a particular way when considering their average or usual behaviours. And it is almost impossible to explain why a particular individual behaves as they do, or predict how they may act in the future. For example, in the early 1960s two American social scientists, Bernard Berelson and Gary Steiner, made an exhaustive survey of more than one thousand scientific studies of human behaviour that they thought had some claim to scientific substantiation. They tried to summarise what this literature told them about human behaviour and came to three conclusions: (i) some do, some don't; (ii) the differences aren't very great; and (iii) it's more complicated than that.

One of the main reasons human behaviour is so complicated is that everything we do is embedded in a tangled mesh of causes. These range from immediate, sometimes unpredictable external demands on our time and attention, to longer-term self-imposed goals and ambitions. Underlying these are ongoing cyclical factors, such as the rhythms of night and day, winter and summer, and the daily routines of sleeping, waking, eating, working and relaxing. Completely external environmental factors also have a noticeable impact, for example, differences in weather and ambient conditions.

And yet, if we seek to provide individuals with insights about their own behaviours, we need to build a system that incorporates some consideration of *why* events happen. This is one of the most important aspects of life.

A natural history of everyday life

Human social behaviour is not the only complex phenomenon that has been scrutinised by scientists. Generations of plant and animal ecologists have struggled to comprehend how complex interdependencies arise within communities of different organisms. In biology, the fundamental unit of study is usually the individual organism. It is the variations between individual organisms, arising from genetic differences and environmental conditions, that furnish the raw material for natural selection. But ecology is different. Although ecology studies biological systems, it is not the individual organism that is the fundamental unit of study in ecology, but the relationship *between* organisms.

Ecology is founded on the methods of natural history, one of the most venerable threads of observational science with roots that can be traced back to Aristotle. Natural history studies organisms in their natural environment and primarily uses observations of how things are rather than experimental modifications.

Although observation has always been important in natural history, in recent years there has been a resurgence of interest, due to the increasing availability of technologies to track and measure in the field. Modern ecologists have looked back into the history of their discipline and found inspiration in the work of the American marine biologist Ed Ricketts (1897-1948), who spent his career studying the marine ecology of the Pacific west coast of the US. There were two key elements in Ricketts's approach. Firstly, his work was based on careful and intense observations of natural phenomena with the aim of understanding *what* happened or *how* it happened. He consciously avoided asking *why* questions and adopted an approach that he called 'is-thinking' – 'which concerns itself primarily not with what should be, or could be, or might be, but rather with what actually *is'*. Secondly, Ricketts always sought to capture as much complexity as possible and then see what emerged from these detailed maps of reality.

These ecological approaches to understanding complexity are a great guide to action in our journey. They imply that we need to bring together multiple types of observation to capture a rounded picture of someone's daily behaviour and lifestyle.

From atoms to profiles

Although atoms are the right means for collecting behavioural data, they are, by design, just tiny elements in an individual's life. To be of real use, the atoms derived from many different devices need to be integrated into structured profiles describing the person's 'life architecture'. The resulting profile is a sort of fingerprint and provides a way to understand the context for particular behaviours. Once we have built these profiles, we can start to predict and anticipate what people may do.

Broadly, there are three types of profile that can be derived from atoms:

– *Lifestyle profiles:* What in total, or on average, are the key parameters that describe an individual's life without regard to behaviour patterns? These include details such as age, gender, body weight, and average daily energy intake and expenditure. Clinicians often try to collect this type of information, though not always successfully. For example, energy intake is notoriously difficult to quantify, as people do not accurately recall what they consume and tend to underestimate the amount. Using atoms, we can make much more accurate assessments of many of these parameters.

— *Character profiles:* What values are important to the individual and how do they keep themselves motivated? This field is properly the realm of psychology, since it comprises thought processes and beliefs that can never be fully or objectively externalised. However, external manifestations of character do exist, and these can be used to infer motivation and preference. There are numerous different approaches to building character profiles and a wide range of frameworks for classifying the personality types within a group or population.

— *Schedule profiles:* What does a typical day look like? And a typical week or weekend? Are there habits and rituals in daily life that even the individual is not fully aware of, but which are nonetheless identifiable parts of their life architecture? Although this might appear the least profound or informative of the three types of profile, in fact the schedule profile proves the single most useful tool. There are few existing approaches to this, showing a clear need for our atom-based system.

The methods we use to create lifestyle, character or schedule profiles are well defined so they can be encoded in detail in a computer program – meaning the resulting profiles are accurate and consistent. Figure 6.1 shows how the atoms captured on different devices can be used to construct a profile. Even without knowing quite how these different data streams relate to each other, we know that they all arise from the behaviour of a single person. This ability to validate to a standard also avoids the need for multiple, costly cross-validation studies between different devices.

Figure 6.2 shows an example of a simple schedule profile derived from Joss's own data. Here the shape of a typical day in Joss's life is shown, splitting his physical activity into three classes of behaviour: standing, sitting or lying, and sleeping. The atom data from seven consecutive days has been used to construct the typical distribution of these three types of behaviour, given hour by hour over an average day. Even without any further insight into Joss's lifestyle we can see it is almost certain that he is asleep between midnight and 5am. His mornings begin at more variable times – there is a forty per cent likelihood he is up at 6am and an eighty per cent chance he is up at 7am.

Creating insights

The general approach to analysing big data is to collect many thousands, perhaps millions, of data points, either in a raw form or synthesised as profiles. Often one type of data or profile is then correlated with another interesting type of data. For example, for a given population it might be interesting to ask how body mass index (BMI)

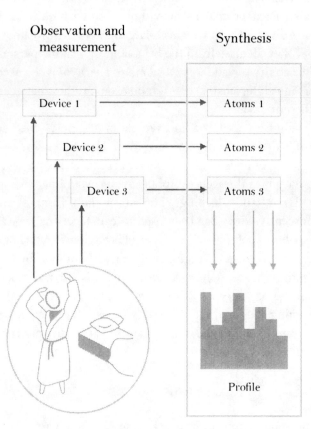

Figure 6.1 *Constructing a profile with atoms. Because multiple devices are capturing data from a single individual, it simplifies the process of synthesising data if all devices use the atom protocol.*

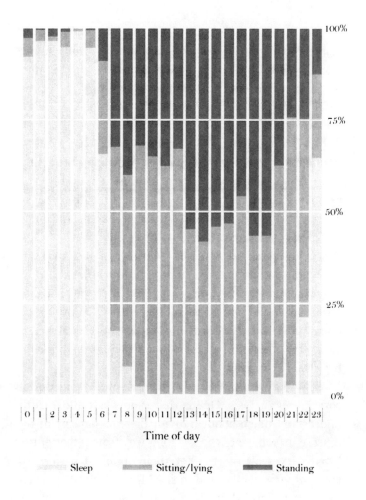

Figure 6.2 A 24-hour profile of physical posture estimated from a wrist-worn accelerometer over a seven-day period. Each hourly block shows the probability that Joss is in one of the three postures.

correlates with the observed pattern of standing time between the hours of 8am and 8pm. This type of analysis would result in a set of data such as that shown in Figure 6.3. Each of the data points in this diagram corresponds to an individual person. Typically data like these are pretty scattered.

A common approach to analysing such data is to try to find a mathematical summary of the whole data set by calculating a line or curve that in some sense 'fits' – for example, the curves (a) and (b) in Figure 6.3. Both these curves fit the observed data to some extent and offer useful ways of thinking about the relationship between BMI and standing time per day. On average it seems that, the higher the amount of standing time per day observed for an individual, the lower their BMI. (Note that this observed *correlation* does not necessarily allow us to understand *causation*, although exploratory analysis of data can often hint at some underlying relationship between two factors.)

There are a number of difficulties with this type of correlation analysis. One is that it does not allow us to understand much about individuals. The data point (c) represents someone who is a long way from either of the fitted curves (a) or (b). Yet their observed pattern of standing time per day and their BMI are perfectly normal for them and probably tell us something interesting about their life. Another issue is that we may generate more and more complex mathematical curves to describe the relationship, leading us to over-fit the data without

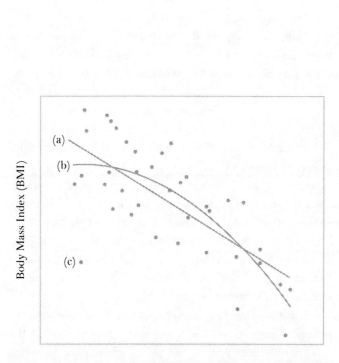

Figure 6.3   A hypothetical plot of Body Mass Index (BMI) versus 'standing time per day' for a population of individuals. The two curves (a) and (b) show relationships that have been found by 'fitting' the observed data with a mathematical model.

creating any real additional insight. Correlation is useful for understanding what populations have done, but in general is not a useful strategy for analysing an individual's daily life or predicting what they may do in the future.

Never invert the data

When we have quantitative data to hand, the most useful question to ask of the observed data set is very often: *compared to what?* When dealing with a specific person, the best strategy is not to analyse their observed profile itself, but to build a coherent narrative of that person's life and from this create a plausible 'mock' profile. This mock profile and the one we have calculated from observations are in the same format (Figure 6.4), making comparison straightforward.

Our task then is to compare the mock profile with the observed profile and see how closely they match. If the quality of fit between the two profiles is not good enough then we need to repeat the process of creating a new narrative and resulting mock profile until an adequate fit is achieved. This allows us, eventually, to arrive at the narrative that best suits the observations. (We should note that the observed profiles include some uncertainty. As Figure 6.2 demonstrates, for each hour in the day there is never certainty about whether Joss is standing, sitting/lying, or sleeping. This means we will never get a mock profile that exactly fits an observed one.)

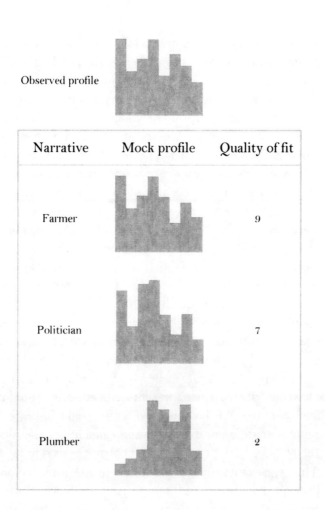

Figure 6.4   An 'observed profile' is compared with three 'mock profiles' we might expect to see for a farmer, politician and plumber. The 'quality of fit' is a way of assessing how close the observed profile is to each of the mock profiles.

This approach is illustrated in Figure 6.4. Here, three potential life narratives and their corresponding mock profiles are shown. Each can be compared with the single observed profile, and a quantitative measure of quality of fit obtained. Based on this set of alternative narratives, we might conclude the observed profile has come from an individual who is a farmer. This insight is not definitive; it may be that, for some as yet unknown reason, their lifestyle merely resembles that of a typical farmer. However, this narrative may well be enough to begin a conversation with the individual, using an engagement strategy based on our working assumption that they are more likely to be a farmer than a plumber or politician.

Although this approach may seem contrived, it is actually what doctors do all the time. With good medical training, it is quite straightforward to predict what signs and symptoms are likely to be seen for a particular disease, or, in other words, to create a mock profile of the symptoms. This is an example of *deductive* thinking. However, when a patient presents with one or more signs and symptoms in combination, these could indicate a number of different diseases – some much more serious than others. In this case, a differential diagnosis is needed. This is an example of *inductive* thinking and is much more difficult than deductive thinking. An experienced doctor tends not to jump to a diagnosis too quickly. Instead they work out what is the best question to ask next, or the next diagnostic test to order. As each new piece of information arrives, the doctor updates their best guess of what the condition is. Each time they do this, their confidence in their diagnosis tends to increase.

The big advantage of inductive thinking compared with deductive thinking is that the conclusions we can draw about unobserved aspects of the system we are interested in can be broader than the observations they are based on. Inductive thinking can generate new insights and knowledge. It can also anticipate and take account of situations where the observed data is incomplete.

Interventions in complex systems

Once we have some understanding of a person's 'life architecture' there may be an opportunity to help them make behavioural changes that will have a positive impact on their health or wellbeing. But we now need to proceed with caution. The key ethical maxim learnt by medical practitioners is: 'Above all, do no harm'. This dictum is particularly useful to bear in mind when we are intervening in a complex system that is poorly understood.

Having made this caveat, our approach paves the way for a wide range of behaviour change solutions:

- *Giving feedback:* In complex systems, particularly those with a strong cyclical component such as daily life, well-timed but relatively small feedback stimuli can have a significant effect on the overall evolution of the behaviour pattern. Remember that data collection and consumer feedback are disconnected, meaning they do not have to be tied up in the same device (see Figure 5.3, Chapter 5). For example, although a mobile smartphone may be the primary data collection device for a particular type of data, this does not limit a service provider to using the same phone to give feedback.

- *Creating dialogue:* Most scientific studies separate out the expensive data collection phase, in which large amounts of data points are captured, from the analysis and intervention phases. In contrast, our atom-based approach allows the data to be used immediately, creating an unfolding and evolving dialogue. The use of atoms also ensures loss of information is minimised if communication is disrupted or incomplete.

- *Visualising insights and patterns:* Humans are dominated by their visual sense. Using emotionally engaging computer visualisations will help individuals to see what the patterns are in their own lives. Despite what data-driven scientists may think, most people are not prepared to devote much time to poring over tables of numbers in search of insights. However, if these data are transformed into images, they can instantly pick up correlations and patterns. With the simplest of visualisations it is possible to help people make sense of everyday life as they currently live it: they can suddenly bridge the gap between what they think they do and what they actually do. The schedule, lifestyle and character profiles described earlier will each have its own range of visualisation styles to help individuals understand their own habits – what is typical or atypical for them. Thanks to atoms, these visualisations can be dynamic and current.

We began from the assumption that although everyday events are atomistic in nature, the overall pattern of daily human behaviour is not susceptible to a reductionist approach. There is no simple way of inferring *why* people do things by looking at a detailed trace of *what* they do. In the holistic and largely cyclical pattern of everyday life it is almost impossible to unravel cause and effect. Asking why someone repeatedly does something, or why they cannot change, never has a simple answer. This is the reason why artificially separating one aspect of daily behaviour from the whole life of an individual cannot create deep and lasting change. There are numerous programs available for dieting, exercise, sleep and mood, but all of them are hampered by their failure to deal with aspects of behaviour that are intrinsically and unknowingly interlinked.

Our approach is different: it is designed from the outset to provide a holistic understanding of daily human behaviour. The system will yield detailed quantitative evidence of what people actually do, when they do it and where. It is on these foundations that we will be able to build detailed and authentic narratives of people's lives and, over time, begin to understand why they behave as they do.

Who 7

A S we start to collect atoms of behaviour from the real world, we can immediately uncover wider information about the population generating those atoms. We could, for instance, view the range of times that individuals wake up in the morning. We might be able to calculate the relative percentages of people who drink tea or coffee first thing. However, if we wanted to discover the preferred beverage of the early risers compared with the later risers, we would need a way to determine which atoms arrived from a particular individual. We would not need any further information about that person and indeed it is likely that they would prefer us not to have further information unless they were receiving some benefit or service in return.

Let us consider what kind of mechanism we would need in order to provide this useful benefit or service. Pursuing our current example, we might want to create a dialogue in which we feed back information about the quality of sleep of our individual – we will call her Sarah. Perhaps Sarah has signed up for a service that gives her objective information about how she is sleeping, with the aim of understanding what she can change in her life to achieve better rest. To send Sarah a message every morning (by SMS text message or email for example), our mechanism needs to link her contact information to the knowledge created from her stream of atoms.

Over time, the ecosystem – Sarah, her actions and atoms, her daily message and the mechanism that allows the message to get to her – will allow Sarah to see the patterns of her behaviour and, maybe, find that reducing her caffeine intake after midday improves her quality of sleep.

The most important point to note here is that Sarah's atoms and the information required to contact her need not be stored in the same place. There are, as we shall see, many benefits in not doing so. On signing up for this service, Sarah would have been made aware by the service provider that her email address was being collected and stored solely for the purpose of providing her with her daily reminder. She would also have been told that her anonymous atoms would be retained for research purposes but not made public. The only way that any information would be released to someone outside the confidentiality agreements (other than following valid legal requests) would be in the form of generalised atoms where the link between distinct individuals was destroyed and any location data suitably generalised. Sarah, for instance, would become one of more than one hundred thirty-year-old females from a given city.

Separating information types

We will now return, in more detail, to the question of the separation of stored atoms and Sarah's contact information. Sarah is much more likely to trust a system where no single organisation has absolute control, especially if there are defined standards monitored externally in a transparent manner. The branded organisation that Sarah has chosen to supply the advice will need to hold her contact information and, if it is a paid service, her payment details. Most consumer-facing businesses have experience of the legal and technical requirements of handling this type of personal information.

However, the collection of atoms and the provision of appropriate advice is a more specialised activity that can best be carried out by a separate organisation which we call a 'data engine'. The benefit for Sarah in this arrangement is that any inadvertent disclosure by a single organisation (for example, as a result of human error, cyber-attack or hacking incident) would not reveal behavioural information relating specifically to her. For the two organisations involved, the service provider and the data engine, this approach makes it easier for them to fulfil their data protection responsibilities to Sarah, provided they follow two principles:

1 There is no duplication of information. The service provider does not collect atoms and the data engine does

not hold personal information that can directly identify the individual;

2 There is a one-way flow of information from the data engine to the service provider.

The data engine holds atoms for many individuals and can do this for many service providers (a business-to-business offering). The relationship between the data engine and the service provider is governed by a contract that defines the commercial relationship, their data protection responsibilities to the individual and adherence to the principles set out above. A direct consequence of these principles is that a single organisation cannot be both a data engine and a service provider.

Devices and the structure of atoms

In Chapter 5, we saw that a device or machine of some description is a necessary intermediary between an individual and any digital system. It is this device that measures, records or infers the atoms of behaviour, which it must then pass on. It is not necessary for the device to 'know' anything about the individual concerned, and so for general information security reasons it is better that it does not. The task of linking atoms from a device to a specific individual then falls to the data engine and, as long as both have unique digital identifiers, this is easily accomplished. It is this mechanism that allows an individual to be served by multiple devices and service providers while maintaining a single stream of atoms. When combined with the Classification of Everyday Living (COEL) it enables a truly holistic view across every domain of daily life.

The nature of the atom-based approach reduces the chance of collecting information that is not directly needed for the provision of a specific service. Atoms of behaviour, like snowflakes, are small and highly structured but also unique, allowing each one to be traced at every stage. This stands in contrast to more traditional data where one fact may be inextricably 'tangled up' with other facts. What is more, the atom structure ensures no personal information is included within the stream of atoms arriving at the data engine. Atoms are truly anonymous: the data engine has never known the identity of an atom's originator and, as long as it is not linked to other data sources, has no method to determine this identity. This is in direct contrast to 'anonymised data', where personal information, and therefore identity, is known to the controlling party and the process of removing it can be flawed or incomplete.

Traditional methods of market and population research use segment labels to characterise individuals within groups. These labels are often based on or derived from information from public records, and say nothing about how individuals actually live their daily lives. A stream of atoms provides an entirely new and much richer resource for comparing groups and individuals within a group. However, to help organisations that need to integrate their findings with more traditional research, the COEL has tools to assign traditional demographic, lifestyle and work segment labels where required.

One particular area of concern for privacy is in the use of location data to provide services. The mobile phone network inherently codes location data as part of its operation and civilian GPS information is now incredibly accurate. The sense of being watched or tracked while on the move provokes strong anxieties. In the next chapter we explore how location information can be coded into an atom and how this can be achieved without compromise to personal identity.

Society and the individual

Considering Sarah's example again, the data engine holds an array of her anonymous atoms, sent from one or more devices. The service provider cannot supply information about Sarah to the data engine but it can provide the linking information between her devices and her digital identifier, as well as request information about her behaviour (such as when she woke up that morning) using her digital identifier. The service provider can then go on to contact Sarah, using the personal information it holds, to give her the service she has signed up to receive.

Here we can see that the power within behavioural data is released when it is linked to other data sources, in this case the personal contact information held by the service provider. The data engine is thus capable of developing significant and meaningful insights about human behaviour at a population level – insights that can be published to deliver social benefits. However, it is imperative that we control the ability of the data engine to link other sources of information to streams of atoms if we are to protect individuals from the risk of unintended identification. As we have already discussed, it is our patterns of behaviour that largely define who we are. Although an individual is not explicitly identified within the stream of atoms, the characteristics of the stream itself can sometimes tell us enough about an individual to identify them when linked to other data sources. In the presence of other such data sources, to discriminate is to identify, and there are numerous examples where individuals have been re-identified from anonymised data such as hospital records and internet searches.

However, we need to find practical ways to exercise this control, since the potential benefits are enormous: large-scale information sets are critical to improving our education, healthcare, transport, and communication systems and to energising free commerce. The ability to put our own behaviour in a wider social context, and understand how we live our lives in comparison to others, is also a powerful motivation for positive behaviour change.

It is clear that we will need to restrict the data engine's right to link, data mine or trade atoms. There are then two very constructive paths to access wider research benefits safely. The first is to replace the anonymous key in every atom with a segment label (age, gender and nationality for example) where there are sufficient numbers within the segment to protect individual identities. This approach provides general information without causal or temporal chains, which can be used by researchers and policymakers. The second path is for the data engine to provide analysis services to answer more specific questions that rely on the preservation of individual data streams. The reports that result from these analyses must confine themselves to trend- and group-based statistics about atoms (meta-atom reports), rather than discussing the atoms themselves, but they can be much more refined than the general information produced by the first approach.

It is the social contract between all parties, ultimately enacted by the data engine, which completes the pathway between the individual and society (see Figure 7.1). Outside of statutory legislation, an open standard with suitable protocols is the only framework in which this trusted and transparent social contract can be achieved. We will discuss open standards in depth in Chapter 10.

Legal and regulatory frameworks

Sarah's example has helped us negotiate the complicated world of personal information while keeping the protection of end-users uppermost. There are, however, a number of further issues to be taken into account: consumers' concerns about privacy; changing social norms;

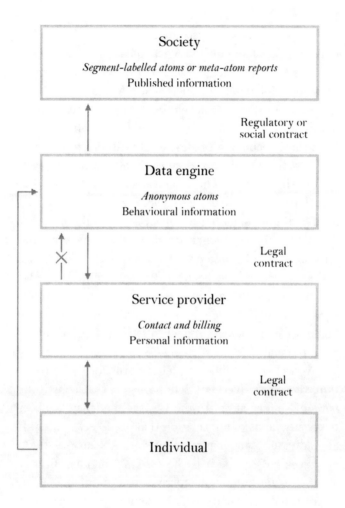

Figure 7.1   The key parties involved in the creation, storage and transaction of atom-based information, with data flows shown on the left and the nature of the contracts on the right.

the attitudes of businesses and non-commercial organisations; and legal and technological questions.

It is within the legal arena that many of the original constructs for describing personal information arose. It is significant that they developed *before* the advent of the information age. The technologies, the benefits derived from these technologies and social attitudes to the risks and benefits have evolved substantially since the early days of the internet. The legal frameworks that regulate personal information in these spaces are adapting and will need continual updating to reflect new social norms.

Although every legal jurisdiction will have its own attitudes, it is the data protection approaches of the US and Europe that have been the most influential. The US has taken a sectorial approach to data protection and legislated only in a few industries, whereas Europe has taken a comprehensive stance, affording legal rights to all individuals.

As well as the differing legal status for data protection, there are divergent definitions of what constitutes personal information. In the US, the tendency has been to specify fields of information that are likely to identify an individual, and hence to prevent the use of these. However, in Europe it is the *potential* for identification that is the determining factor and so, as more linked data is collected and analysed, the definition covers ever-widening classes of information. In contrast, the ability to mine data and data patterns for other identifiers will soon leave a list of specific fields (as preferred in the US) redundant. Despite these substantial differences in approach, the underlying principles are reasonably consistent. An individual should expect any organisation that collects or holds personal information about them to:

1 Be *accountable* to the principles.
2 Be *transparent* about its practices.
3 Ensure the information is *accurate*.
4 Treat the individual's information *securely*.
5 Allow the individual *control* over their information.
6 *Limit* information to that needed for the agreed purpose.

It is partly in disregarding this last point that the problem of big data arises. The collection of large quantities of data without clearly stated aims gives rise to the very real possibility that a moral conflict will arise after a data collection contract has been agreed. Merely the perceived risk of this moral conflict, in combination with the knowledge that the power (i.e. the data) rests with the larger party, will inevitably lead to concerns about Big Brother.

For example, let us consider a situation where an insurance company collects location data about a vehicle, with the driver's consent, to ensure it has an accurate assessment of risk. If this location data is sufficiently detailed (both in the frequency and accuracy of measurement) then it will be possible to deduce the velocity of the car. This additional information is not needed for the stated purpose of knowing the car's location, but it is difficult for a business or technologist to throw away data that appears to come for free. It is at a later stage (after the contract for the collection of data has been agreed) that the insurance company becomes aware that the driver substantially exceeds the speed limit past a school every morning on his way to work. The moral dilemma here is easy to see and would take several chapters for us to explore. However, it illustrates just how vital it is that data protection regulations serve the interests of both individuals and businesses.

Behavioural data in context

Lastly, in Figure 7.2, we show how the degree of sensitivity of personal information can vary according to the different domains where it is used. As the internet developed, online companies were free to use information of low sensitivity as a commercial asset to build their businesses. Simultaneously, governments and regulated industries saw the potential of these same technologies to reduce cost and increase productivity while processing highly sensitive information, such as health records, that would remain confidential.

The quantity and sensitivity of general behavioural information now being collected, in combination with increasing analytical capability, is creating a growing demand for some areas to be more carefully controlled. There is also rising pressure to relax heavily regulated areas to allow more research for social benefit. Both

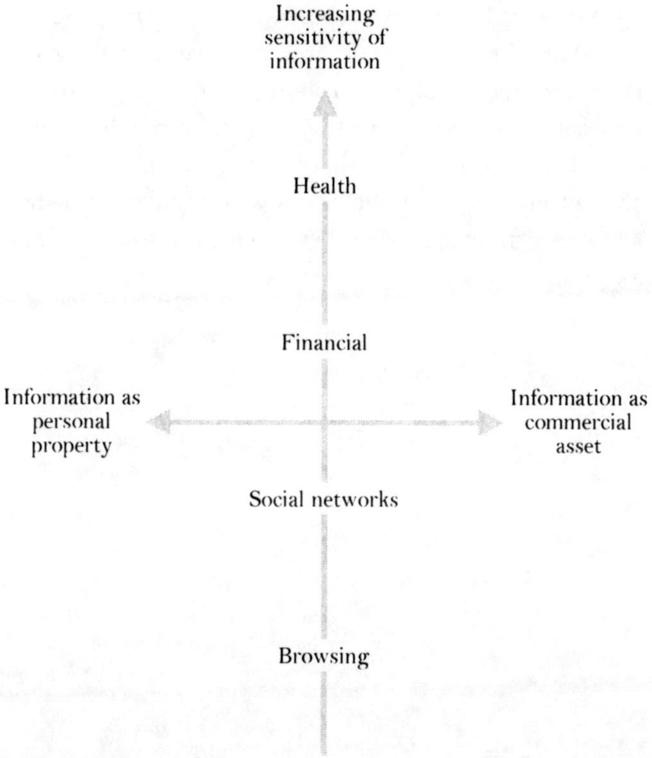

Who

Figure 7.2 A mapping of the personal information space showing a number of different domains where information can be interpreted as a commercial asset or personal property.

lobbies will be looking for new models and we believe an open standard provides the best opportunity for progress.

Many look to regulators to drive this progress. However, the regulators' role is essentially negative – they can only restrict unwanted practices, not devise new ones. It is therefore vital that businesses innovate in this area, prompting fresh dialogues with regulatory bodies. The consortium approach we advocate is the ideal vehicle for doing this without undue risk.

Who

Where 8

IMAGINE finding yourself in the middle of São Paolo in Brazil, or out on the Nullarbor Plain in Southern Australia. Although the surroundings might be unfamiliar, you could use your senses to build up a picture of the immediate environment and, before long, gain some understanding of your predicament. However, although in both locations you could get a rudimentary understanding of what was going on around you, this would not help you work out where on the surface of the Earth you were. You would be completely lost.

Although humans have inherited from primates the ability to understand the three-dimensional space of our immediate surroundings, we have not evolved any specific way of sensing where we are geographically. Due to the action of gravity we can always sense the difference between up and down. In addition, over a period of twenty-four hours we can roughly identify north, south, east and west by observing sunrise and sunset. And that is as far as we can go. The simple question 'Where am I?' cannot be answered by a human without using specialised technology, except when in familiar surroundings.

When at home, I don't need any technology to tell me where I am. The 'Where am I?' question can therefore be answered in two distinct ways: *home* is both a precise latitude and longitude on the Earth's surface and a human space in which I live and am most content.

Before people had developed maps or the means to determine longitude, latitude and direction, they travelled long distances on foot, using well-worn paths. These paths do not need a complete understanding of every square metre of a landscape; instead they use natural features to help orientate the traveller. Streams and rivers provide markers and a sense of direction: upstream and downstream. Valleys enable transitions from higher to lower ground, while rock formations, mountain peaks, individual ponds and trees all offer a means to navigate.

Walking helps humans create a detailed understanding of macroscopic places as experienced through journeys. Transit patterns between different places have given us rich and complex pathways through the world and repetitive use of these paths has created important routes that have lasted much longer than any individual human lifespan. Some of these networks remain important routes for travellers and pilgrims, although many historically important paths and tracks do not appear on maps.

Over the past five hundred years, people have developed a range of important navigation technologies that help us pinpoint our location. These include maps (an analytical representation of the physical layout of the Earth's surface), longitude and latitude positioning devices (to tell us where we are in east-west and north-south directions) and the compass (to keep our bearings on a journey). Nowadays longitude, latitude and direction can be determined with a GPS device, while interactive maps such as that used on a car satellite navigation system show our movements in real time. However, in the past it required considerable technical skills and equipment to discover your location and direction. A seafaring navigator of two hundred and fifty years ago needed a full set of detailed charts and maps, a sextant to determine latitude, a chronometer to determine longitude and a compass to determine direction. Our own familiarity with precise navigation and long-distance travel is a very recent addition to human experience.

Navigation technology has helped enrich human experience and made a person's birthplace or home country less of a constraint on their experience of the wider world. Other technologies have also loosened the dominance of location in human lives: transport (horse, train, car, boat, plane), long-distance and asynchronous communications (mail, telephone, radio, TV, SMS text), recorded images (sketches, paintings, photographs, movies) and the internet. Yet despite the massive changes that these technologies have brought, the fact remains that everything that happens in our daily lives must happen *somewhere*. Location is perhaps the single most powerful way we have to contextualise an event.

Imagine wearing a small camera around your neck that has been set up to automatically take and record a picture once every thirty seconds. After a week we could review the array of images the device had recorded, and this would give us a stop-frame film of the environments we had encountered in the intervening days. This movie would reflect both where we have been and what we have been doing. The images could help us reconstruct a narrative of our activities because it recorded our location. From this spatial record we could create a reasonably accurate behavioural record. Where we have been is not exactly the same as what we have been doing. But it certainly is a clue.

Where we do things is almost as important in our understanding of our everyday lives as *what* we do. The places we frequent and those we invest with particular meanings – home, work, school – provide vital contextual information for understanding our behaviour. This context helps structure our social groupings, from the small-scale composition of our family units to the large-scale make-up of villages, towns, cities and countries.

Space and place

Long before the development of the modern concepts of analytical geography, humans had both sacred and pro-fane places in their lives. A place is not simply a particular location on the surface of the Earth. We have to distin-guish between *space*, the technical description of a location in a well-defined analytical representation, and *place*, a location with a human description. The place where an event happened is one of the most basic elements to include in a useful description of a particular event or pattern of events, and our journeys through and between these places reveal some of the most habitual aspects of our daily activities.

Place as a concept has broad implications for our ordinary lives and for our rituals and routines. To define place in a way that is useful to include in an atom of behaviour, it helps to think of it as having three qualities: location, material form and meaning.

All locations on the surface of the Earth can be identified uniquely using latitude and longitude. Once a location is defined as a specific place, then we can make comparisons between where we are (here) and another place (there). For example, imagine standing at the front door of your home. This place is unique and can be described using an internationally recognised set of co-ordinates of longitude and latitude. So can every other place you might imagine. Once we make a comparison between two different places we can also describe distance, either roughly (near, far, very far), by physical distance (three kilometres) or by temporal distance (thirty minutes' drive).

Although all places have unique locations they are not all the same size. We can identify places that have real meaning for us but are tiny on the scale of the Earth, such as the exact corner of one of our rooms that gets the light through the window first thing on a summer morning. But places can also be much larger and have real physical extent – our home, our neighbourhood, our town or country.

All places have a definite physical form. A place includes an assembly of things that, if described in sufficient detail, is unique to itself. Even places that outwardly resemble each other cannot be identical. All outlets from a particular fast-food chain look superficially the same, but on closer inspection they reveal their uniqueness. The physical forms that define a place can be natural in origin, as seen at Yosemite National Park in California; semi-natural, including significant human modification, as seen in the distinctive pastoral landscape of England's South Downs; or entirely man-made, such as the Eiffel Tower in Paris.

The physicality of a particular place is perceived with our five senses: sight, sound, taste, smell and touch. These sensory impressions may linger with us for many years, the smell of chalk recalling a childhood classroom, for example. Specific places can have a powerful impact on us, imprinting themselves in our memory so that even decades later we can remember them in detail, and may experience strong emotions if we revisit them.

Although every location on Earth can be identified and has a unique physicality, unless it has been accorded significance by humans it is not strictly speaking a *place*. If a place is denuded of its human significance, it is merely a location. A place is not a set of co-ordinates, an anodyne setting in which something happens, it is a *place* because it represents some key human activity or meaning. A family's dining room is created not just by the type of objects typically found there within a particular culture, but also crucially by the human ritual involved in sharing a meal.

These three components – location, material form and meaning – are what transform a technically recorded *location* into a *place* that is relevant to us. This begins to explain why it is not helpful to give someone a map of where they have been recently, constructed solely from GPS co-ordinates. These maps do not show any places of personal interest and therefore fail to engage an individual's emotions.

The three facets need to be bundled together when considering the role of place in an atom of behaviour. Place is a rich source of context about human activity, raw location data is not. The Classification of Everyday Living reflects this aspect of daily human activity in its ability to use everyday place names such as school, mosque and so on, as well as longitude and latitude. This approach also protects key private information about a consumer. There is already unease among consumers about the tracking of their online browsing behaviour and this will deepen if their precise locations start to be tracked. Exactly where an individual is at any one time, or where they can reliably be predicted to be at a future given moment, is perhaps the most sensitive of the different types of behavioural data a service provider can record.

Mobility patterns

Humans are mobile by nature. The global radiation of our ancestors from Africa shows that, even on foot, humans can cover enormous distances given enough time. We can categorise human mobility into three broad types: migration, daily place-to-place movement and daily room-to-room movements within a defined place such as the home or workplace.

Migration is the temporary or permanent movement of an individual or group from one location to another. Implicit in this definition is the idea that the two locations are sufficiently distant that migration does not happen on a regular basis. Often migration will be driven by significant social, political or economic forces such as famine or conflict. It will almost always have a significant impact on the migrants themselves, as well as on those they leave behind and those with whom they live at the new location. Since the 1880s, social scientists have been formulating 'laws of migration' to explain observed migration patterns. Recently, the advent of mobile phone records and much more detailed census data have enabled the development of more robust and generally applicable models of migration.

Imagine following an individual around during their daily life, recording exactly where they went. The chances are, the pattern will be dominated by a sequence of short trips between locations. Although large-scale and long-term mobility patterns have been studied for many decades, it is only recently that it has become possible to record people's trips in sufficient detail to understand the daily mobility patterns of individuals. This information can be obtained from mobile phone data, GPS records from smartphones and from geo-located social media records. A number of recent studies using this type of data reveal three key features of our daily mobility patterns which are worth considering in more detail:

- There are a very limited number of daily mobility patterns, or 'motifs'.
- The number of times we visit different places is very unequal.
- The correlation between our spatial and temporal behaviours means there is a long-term predictability about where we might be found.

Motifs

Imagine we had collected a detailed record of the trips made by an individual over a number of weeks. If we ignored the quantitative aspects of their movement, such as the distance travelled or the time taken, and simply recorded the pattern of the places they had visited, we would end up with a motif that represented their mobility pattern.

Figure 8.1 shows some of the most common mobility motifs identified in a recent study. Each motif shows the discrete places that have been visited as a circle, with arrows showing the direction of trips made between these places. This representation ignores many of the details of real trips to emphasise the daily pattern of movement for an individual. For example, motif 2 might be a day that begins at home with a journey to another place, such as school, and back home again (and this pattern may be repeated in the same day). What is surprising is that only seventeen different motifs are needed to account for more than ninety per cent of the daily trips analysed. Furthermore, of these seventeen motifs, only a small number covers the vast majority of daily trip patterns. More than fifty per cent of observed daily motifs are of type 1, 2 or 3. Most of us have simple daily patterns of mobility. Some of these motifs are somewhat unexpected, for example motif 4 begins at one of the places and involves a circular journey to take the person back to the start. It appears that individuals habitually follow a characteristic motif, with a stability lasting over a period of months.

The same type of data that is used to create motifs can also be used to describe the frequency with which an individual visits a set of specific places. Unsurprisingly, observed visitation patterns are not uniform. Not all of the places we commonly go are visited equally often.

For example, if we found over a number of days that a person had visited a total of twelve different places, we could add up the number of trips to each location. Suppose they made a total of one hundred visits to the twelve places. We would expect the observed visitation distribution to look something like Figure 8.2. In this example, the most visited place, which for the majority of people

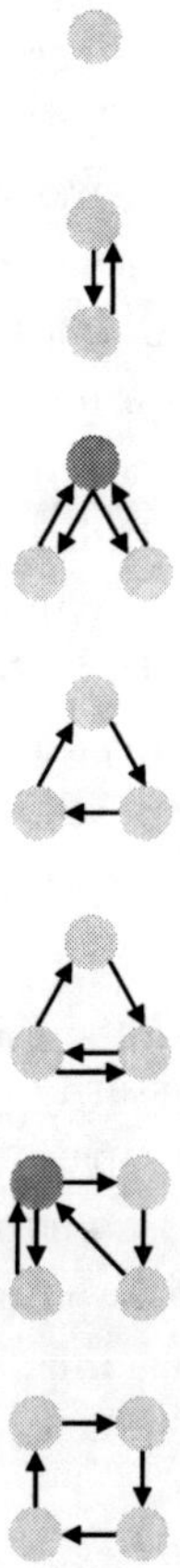

Where

	Motif	Frequency
	1	10%
	2	30%
	3	10%
	4	10%
	5	2%
	6	7%
	7	5%

Figure 8.1 Common daily mobility motifs. The circles represent a discrete place that an individual has visited. The arrows indicate the direction of travel between each of these places and the darker circles represent a place that acts as a hub.

would probably be home, is visited thirty-two times. The second most visited place, which for many people would be work, is visited roughly half as many times as the most visited, in this example sixteen times. The third most visited would be roughly one-third of the most visited, in this example about ten times. And so on.

To further illustrate the unevenness of how people spend their time at different places, Figure 8.3 shows the distribution of Joss's mobile phone location over a six-month period. Joss spends about six nights per month away from home and is highly mobile, travelling around the UK for work, family visits and holidays. However, he still spends more time at home than anywhere else: his home patch covers sixty-three per cent of all points, while only eighteen locations account for ninety-four per cent of the points.

Note that the number of times we frequent each commonly visited place does not necessarily reflect how emotionally significant a place is. Perhaps in the course of a week there is only a single visit to a church or elderly relative's home, but that visit may have a special meaning in our lives and the place may be more important to us than the simple frequency plot would indicate.

Home

Home is not just a physical place: it is fundamental to our wellbeing. To be without a place of one's own is deeply unsettling. Even though for nomadic peoples this home

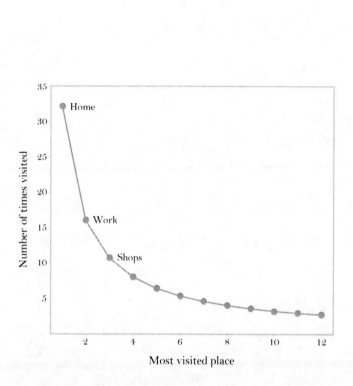

Figure 8.2 The typical frequency distribution for an individual's pattern of visits to different places during the course of their weekly or monthly activities.

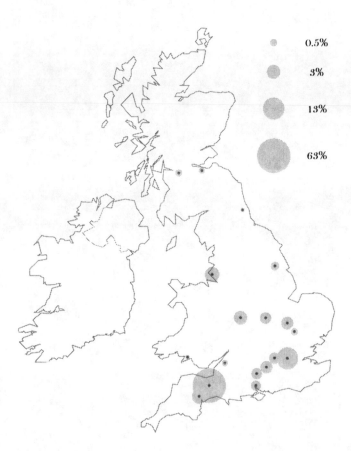

Figure 8.3   The spatial distribution of Joss's locations across the UK over a six-month period. The location data is resolved to the nearest mobile phone cell.

may be mobile – a gypsy caravan, yurt or teepee – it is still a fixed reference point psychologically. Home is special in everyday life because it is generally where we begin the day and where we return after work or the journeys we make to other important places in our daily routine (school, shops, workplace).

Home provides us with a sense of familiarity, protection and privacy. We often put considerable effort into decorating our homes to reflect our personality and project an image to the outside world. Home also acts as a storage place for objects that symbolise and preserve our life histories: trophies from our sports teams, gifts from our children or grandchildren, family photographs. Home making is not just housework.

For homes with more than one room, different rooms will be used for different things, and some rooms will be much more heavily used than others:

1 A room for food preparation and cooking.
2 A room for eating and sharing food.
3 A room for family living, socialising and entertainment.
4 A room for sleeping.
5 A room for grooming and cleaning.
6 A room for toilet activity.

In different cultures and socio-economic groups the functions described above may be merged together: perhaps one room would be used for both preparing and eating food. Within each of these rooms there may be even more detailed and identifiable places. The living room may include *my favourite chair*, *the spot where my child usually sits* and so on.

The detailed mapping of daily human activities within the home and in individual rooms is in its infancy. Wearable digital devices with very high spatial resolution of less than about ten centimetres would be required to record which rooms an individual habitually visited. Even though today's electronic GPS devices provide impressive spatial resolution they are still unable to record human locations within a home accurately. Part of this is a technical issue, but a bigger factor is that within the home different rooms may be much more sharply defined from an emotional point of view than from a physical one. In a real home the bathroom and bedroom may be physically separated only by a wall six to twelve centimetres thick, yet these two rooms will have completely different meanings for their occupant.

Recently a set of technical developments have come together to create the concept of the Internet of Things, as described in Chapter 5. The basic idea is that as the cost and size of sensors and communications technologies reduce, it will be possible to distribute millions of small, low-cost sensors into the world, each connected to the internet. When combined with a suitable data collection protocol to record device ID, location, time and event type, this network of sensors could create robust data streams of low information content. The obvious way to use these would be as a set of sensors of human activity in the home.

It is likely that an individual's pattern of use of the different rooms within their home will not be uniform. In the same way that we find not all of the places we commonly go are visited equally often, it is likely that not all of the rooms in our homes are visited the same number of times.

As the pattern of visits to different rooms becomes better understood, a need to analyse the micro detail of our homes will emerge. Over the past fifteen years, the sociologist John Grady has used a simple technique with students in his visual sociology class to help them map and describe the objects kept in their personal rooms. He gave them a simple diagram of an exploded view of a rectangular room, and asked them to make an illustrated map of their belongings. He also asked them to list three of their favourite things, photograph them and write accounts of how they had obtained the objects and why they were special.

Grady now has a collection of more than two hundred and fifty of these visual maps and has noticed three things: (i) many of the possessions were valued because of their connections to other people, particularly the students' families; (ii) few of the most prized possessions had any appreciable commercial value or were status assets; and (iii) the students put a large amount of energy and attention to detail into this piece of work, even though it was a small contribution to their overall grade. Grady's study makes it clear that, even for students in temporary accommodation with few belongings, it is the emotional dimension of possessions that helps to create a private place.

Long-term predictability of location

If the mobility pattern of an individual is combined with other important contextual information, such as the day of the week, then it becomes possible to predict with good accuracy not only where the individual may be next week but also far into the future. This level of predictability is unsurprising. It is the outward manifestation of the fact that, for many people, daily patterns of activity are intimately connected with where they are. There is also a strong correlation between different types of event and places. Sometimes knowing the *type* of event that has happened will help to identify the type of place where it occurred. Conversely, sometimes knowing the *place* where something has happened will help to identify the type of event it is. Knowing somebody is using a computer at their workplace is a good clue that they are working.

The ability for an individual to anticipate their likely location at any specific time is a source of significant insight. It can help to plan more successful behaviour change or predict when we are most prone to failure. An understanding about place can trigger specific elements within a plan, such as using the knowledge that someone is at work to give reminders to break up long periods of sitting. It can also be used to enhance a piece of advice, for example by suggesting a route for a short walk at lunchtime.

Our personal histories and geographies interact, everywhere and all the time. If location, event type and time are all available, as they are in an atom, it becomes possible to understand in real detail the daily life of an individual.

Data to Life 9

THE ability to classify knowledge is a fundamental step in the development of science. It gives equal access to those inside and outside the area of study, while providing a logical framework to acquire new knowledge. The structure of the Classification of Everyday Living (COEL) transcends language: it is universally applicable.

When helping people adjust their ingrained habits, it is essential to measure behaviours at the same level of detail as the changes you want to support (one more apple or one less cigarette). With atoms, we can construct profiles that allow us to predict and anticipate behaviours, rather than merely react to them.

Figure 9.1 shows the power of atoms to clarify and visualise our activity. The first set of plots shows raw data gathered by the movement sensor on Joss's wrist-worn device. The second set covers the same period, but this time uses atoms to focus on Joss's sleeping and activity levels. The raw data resembles a hospital chart, where it might well be appropriate to present the findings in their entirety, so they can be considered by a trained professional. However, when dealing with everyday behaviours, we need only enough information to help us understand

Data to Life

Saturday

Sunday

Monday

Tuesday

Wednesday

Thursday

Friday

04:00   12:00   20:00   04:00

Data to Life

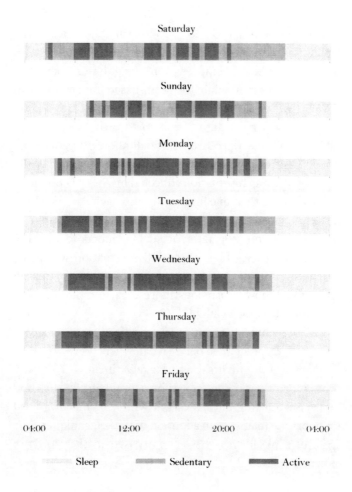

Figure 9.1 *The same week of activity shown as raw data (left) and as atoms (right).*

the wider picture, provided we are still able to detect the smallest meaningful changes. The raw data stream illustrated here requires ten megabytes per day (10,000,000 bytes) while the atom-based information can be sent with just five kilobytes (5,000 bytes).

Engineering a digital system to process atoms has led us to define a protocol that will allow all people and machines to record behaviours in the same way. Over the past two years, information collected by Joss has provided us with more than one hundred atoms each day. We have also processed thousands of other atoms from a range of individuals, monitoring areas such as mood, food, drink and internet use. The sources of information will grow with time as devices acquire new sensors, increased processing power provides more local intelligence and new apps become available.

The technical structure of a digitised atom of behaviour with its predefined fields (what, when, how, why, who and where) is universal and largely timeless. Not only does it provide a common platform for Quantified Self and exercise-tracking enthusiasts, it will also encourage and support the mainstream adoption of these techniques. The role of a platform is best summarised in the words of Larry Keeley:

'A platform is an integrated offering that creates a unique and holistic customer experience only loosely controlled by the platform owner. It is usually supported by proprietary technologies, and typically characterised by interdependent products and services provided through a network of business partners.'

It is this approach to a platform, in combination with a commitment to interoperability, versatility and humanising technology, that has driven the creation of Coelition.

Designing an engaging service

The digital service industry is already adept at planning and executing seamless interactions between consumers and brands. But personalisation is the single most critical element when building an emotionally engaging dialogue with a consumer, and until now this has been lacking. Although responsibility for the consumer interface rests largely with service providers, it is the Coelition platform that enables the depth of personalisation required. This attention to detail gives automated systems the ability to provide advice that is as tailored and persuasive as words of encouragement from a personal trainer or lifestyle coach.

To be fully effective, digital service providers should follow these principles when giving advice:

– *Content must be set in the context of my daily life.* Giving consumers information or guidance that is correct but out of context reduces the chance that they will notice, accept or act upon it.

- *Show me that you know me.* Once a consumer has disclosed personal information (name, age, gender or anything else) to a system, they expect that information to be used sensitively to tailor the product or service they are buying.
- *Empower through strengths.* Systems should find ways to tap into the consumer's strong positive habits and rituals as a way to harness energy for change.
- *Small steps add up to a big impact.* Breaking down big changes into achievable steps is essential if lasting shifts are to occur.

For us as consumers the promise of a better service, tailored to our individual needs, is always a compelling proposition, but the impact goes much further. A platform approach gives disparate apps and devices the ability to talk to each other safely: your cycling route tracker will automatically update your calorie counter or the service schedule at the bike shop.

These machine-to-machine communications are not conveyed as some remote digital transaction, but as atoms that have meaning to you. Each one is marked with your unique digital identifier, giving you direct benefits from it as well as control over its use.

We know our digital information is valuable – but it is so mainly when pooled with the information of others. If we wish to realise this potential as individuals, we need to work within communities to create aggregated information with a tradable value. This might be in the form of lobbying for marginal groups, local communities or sets of people with similar health issues. So, for example, a group of diabetes sufferers might record their lifestyles and treatment regimens to create a dataset that could be used as a research tool by a drug development company, in return for health services.

In all cases, consumer brands working within the Coelition platform will be well placed to help individuals, communities and organisations to understand everyday life and make the behavioural changes that are important to them.

Behaviour change for health

Before the Second World War, communicable diseases were the main global cause of death, and average life expectancy was low because of them. Over the past sixty years, medical and public health research has led to mass vaccination programmes, the development of antibiotics, and improvements in basic nutrition and hygiene. Together, these have led to a decrease in deaths from communicable diseases.

However, during the same period, an increasing number of deaths from non-communicable diseases (NCDs) has become a major problem for industrialised countries and now for developing countries as well. By 2020, it is predicted that NCDs will cause seven out of ten deaths in developing countries. The World Health Organization reports that: 'Non-communicable diseases are the leading cause of death globally, killing more people each year than all other causes combined'.

Many of these NCDs are directly linked to the rapid changes the world has experienced over the past half century: ageing populations, rapid urbanisation and unhealthy lifestyles. The four biggest risk factors for NCDs are all related to lifestyle choices: tobacco use, an unhealthy diet, too little physical activity and alcohol consumption. All these risk factors work in the same way: people make hundreds of small, poor choices each day, and the cumulative effect over months, years and decades is cardiovascular disease, cancer, type II diabetes and chronic lung disease. None of these can be eradicated by a vaccine and there is no 'magic bullet' pharmaceutical. They are all simply and unequivocally reduced or stopped when individuals make positive lifestyle changes: curbing fat and sugar intake, reducing or quitting smoking, increasing physical activity, and cutting alcohol consumption.

And yet resistance to change remains profound. Although the dangers of obesity, smoking, poor diet, insufficient exercise and alcohol abuse are well known, it is extremely difficult to alter the behaviour of individuals permanently and even more challenging to instil lasting change across a whole society, as governments well know. The relative success of different approaches to changing behaviour has been well documented in the fields of health and personal care. For example, when it comes to helping an individual establish a healthy tooth-brushing regime, it has been found that forming routine self-regulation is more successful than motivation, which in turn is more successful than education. Even when armed with such findings, translating theoretical psychological constructs into an effective set of interventions that will really help people alter their habits is a complex task. However, five working principles will be useful:

1	Awareness and attitude	(Make it understood)
2	Self and society	(Make it desirable)
3	Easy and difficult	(Make it easy)
4	Proof and reward	(Make it rewarding)
5	Permanence	(Make it a habit)

These principles are not in order of priority; they can work independently as well as in combination, and the more that are applied within a single situation the more likely it is that an individual will successfully adopt a new behaviour.

It is clear that promoting and enabling positive lifestyle changes through the delivery of trusted, branded services is a powerful weapon in the battle against NCDs. There are already some good local government and commercial initiatives in the field of digital health innovation, but they lack a common infrastructure to give them a truly global impact. Coelition can provide this link. Whatever the exact nature of the service, it will contain a variation of the assess-plan-intervene-assess cycle. The core strength of the Coelition standard is its ability to define the *assess* part of the cycle, reliably and consistently.

Today, the slowing rate of new drug discovery is ushering in an age where medicines will be most effective when used in conjunction with behaviour change techniques. To see the truth of this, you only have to look at smoking cessation programmes, where cognitive behavioural therapy or nicotine replacement on their own have mixed results, while in combination prove highly effective. Genetic analysis promised to fill the gap left by fewer new drugs coming onto the market, but it too has somewhat stalled, due to the growing understanding of the vital role played by lifestyle in health outcomes. To make sense of this complex picture we will need tools that can quantify lifestyle with the precision we have come to expect from the rest of modern science.

Finally, ageing populations in almost all societies are forcing governments and health providers globally to re-evaluate their models of healthcare provision. These new models will rely on more distributed, community-based tools, where again Coelition has a significant role to play.

Environmental sustainability

As well as facing challenges to human health, we are also confronting serious challenges to the health of our planet. The social changes of the past two hundred years have accelerated in the last fifty, placing fundamental pressures on our environment. As with unhealthy lifestyles, no immediate damage is seen from any single action, but

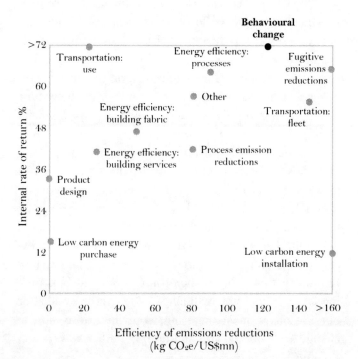

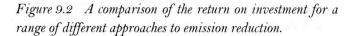

Figure 9.2 A comparison of the return on investment for a range of different approaches to emission reduction.

collectively these actions are slowly killing us all. Again, behaviour change is one of the most effective solutions.

The Carbon Disclosure Project analysed more than two hundred of the world's largest companies to identify how they were reducing their carbon emissions. It examined both the environmental impact of every dollar spent and the financial rate of return on the investment. Figure 9.2 is a summary of its results, with behaviour change being highly attractive both for its financial impact and for the environment.

Our individual consumption of water, energy and materials is determined by the tiny decisions we make every day – mostly out of habit rather than through any rational process. The ability to determine these daily patterns across large numbers of people will be critical in designing behaviour change strategies to ease the pressure on our planet's resources. The growing potential of the Internet of Things, when coupled with the humanising layer of Coelition, will allow us to measure and alter our use of the most important contributors to climate change such as cars, air conditioning units, showers and washing machines.

Moving forward

We are not short on data. It is the imagination to tackle our problems in a different way that is currently lacking. Coelition is only one part of a number of potential solutions, but we have imagined it, engineered it and our prototype system is up and running. The COEL will continue to grow, covering other languages and cultures in increasing depth. The atom-based approach provides a unique framework to explore the interplay of our behaviours and perception of time. It will allow a much better understanding of how interventions (both medical and behavioural) can be phased within our daily cycles for maximum effect.

With an accessible protocol in place, the tools for collecting atoms can multiply, fuelling ever more sophisticated analytical approaches. It is the development of strategies to communicate complicated and challenging information to consumers that may prove to be the most significant hurdle for behaviour change systems in general. Designers, behaviour change specialists, brand owners and specialist professionals (such as medical practitioners) must work together in an iterative cycle to create visualisations, interactions and experiences that can motivate personal change.

Data from digital technologies is weaving its way ever more closely into our daily lives. Starting with the simplest atoms of behaviour, we have set a path through the issues and opportunities of this advancing revolution. This book has described our exploration of everyday life – reducing our complex daily patterns to data, processing the data and finally bringing that data to life.

Coelition, and the Coelition standard, will allow our partners to design, test and implement their ideas more swiftly and thoroughly than ever before. We set out the structures that enable this in the next chapter.

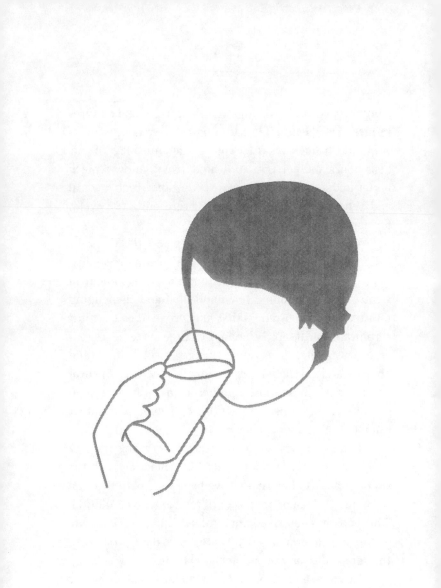

Coelition 10

'The most profound technologies are those that disappear. They weave themselves into the fabric of everyday life until they are indistinguishable from it.'

THIS statement, taken from a 1991 article in *Scientific American* titled 'The Computer for the 21st Century', signalled the start of what computer scientist Mark Weiser saw as a third wave in computing technology: ubiquitous computing.

We now live in the era imagined by Weiser. Tiny computers are embedded into many of the objects we interact with in our daily lives and are wirelessly connected over longer and longer distances. Furthermore, the mobile phones we habitually carry with us are, in fact, powerful computing and measuring devices, capable of wireless connection to hundreds of other devices as well as to the internet.

Although modern hardware and software technologies and machine-to-machine connection protocols are necessary to make ubiquitous computing a reality, until now they have lacked the final, humanising layer that would make them truly 'disappear' and 'weave themselves into the fabric of everyday life'. That humanising layer can now be added. Previous chapters have shown that pervasive digital data collection is possible. We can define in detail the atoms of behaviour of everyday life. We can record these atoms when a machine infers they have occurred, by analysing a human interaction with its sensors or other inputs, and transmit them either immediately or at a time convenient to the machine. All that is required now is a workable standard for collecting these atoms from a wide range of digital technologies. This standard, which we have named Coelition, is described in this chapter.

Established in the UK in 2013, Coelition is a not-for-profit organisation. It is a consortium of leading international businesses that want to measure and understand everyday living in the digital age. Coelition unleashes the potential for real benefits from modern computing power, pervasive sensing and cloud networks, while avoiding the headaches of big data and the worries of Big Brother. Its primary role is to develop, publish and manage an open standard that will allow real behaviour to be measured against an international benchmark and individual results to be aggregated – all with data protection designed in.

Systems that implement the Coelition standard will enable a direct dialogue between consumers and the brands they trust, helping both parties to make sense of everyday behaviour. Their findings will provide insights that are accurate, anonymous and actionable. Coelition will empower individuals, organisations and communities to effect beneficial, lasting and profound behaviour change.

The Coelition standard consists of three main elements:

1 The Classification of Everyday Living (COEL) acts as a powerful library that allows everyday human behaviours to be identified and encoded with unprecedented detail and accuracy. The value of this asset largely derives from the rigorous approach we have taken to ensuring it is genuinely holistic. We have not found another taxonomy of everyday activities that comes close to either the depth or breadth of the COEL.

2 The Atom Protocol encodes the what, when, how, why, who and where of everyday events into compact and richly detailed atoms, and provides a set of methods for sending these atoms from any application or device via networks of any bandwidth.

3 The Identity Authority (IDA) transparently manages identity protection for those using the standard. The IDA is responsible for issuing unique digital identifiers for every consumer using the system. These are created and co-ordinated centrally to ensure consumers can engage with multiple service providers using a single identity, if they so choose. This digital identifier is held securely by the service provider and ensures that atoms can be linked only with an individual's permission. Each request for a digital identifier constitutes a legal agreement between

the service provider and Coelition, covering the correct and responsible use of that identifier and associated personal information in perpetuity.

The Coelition standard has been designed to facilitate a wide range of business-to-business (B2B) and business-to-consumer (B2C) services, by incorporating the engineering design principles set out in Chapter 5. Two further tenets are central to the standard:

— *Systems should be modular, flexible and non-monolithic.* New device platforms are constantly being developed. By using a device agnostic standard, companies that make devices and capture data from consumers will obtain more value from the data.
— *Manage information not data.* Data is bandwidth-hungry, messy, context-specific and has a big storage requirement. Information requires lower bandwidth, is less context-specific and requires less storage space.

Any brand collecting personalised information to bring about behaviour change will achieve the best outcomes if it implements the service design principles described in Chapter 9. These are not mandated by the standard, but it has been designed to support them.

The Coelition standard in context

The Coelition standard is designed to help organisations map the broad range of everyday consumer activities that sit between purely commercial behaviour (such as online purchasing), self-projection behaviour (such as status updates on Facebook and other social media) and medical-related behaviour (such as clinical measurements, including blood and cholesterol tests).

Currently, at one end of the scale we have purely commercial online behaviour, which is, by definition, confidential and proprietary to the retailer making the sale. A prime example is Amazon, which records in detail the online activity of its shoppers and uses this data to run predictive recommender algorithms that offer suggestions for future purchases. 'People like you have bought x, y or z...' is a powerful way to exploit the insights a retailer has obtained from its customers' shopping patterns. These retailers do not subscribe to a common, standardised means for classifying online behaviours, nor do they need a more widely defined standard for everyday behaviours that happen away from consumers' shopping activity.

At the other end of the scale is the medical profession, which is tightly controlled. Here the individual is not a consumer but a patient, and under these conditions there are significant ethical and regulatory restrictions on the collection and use of data. These restrictions are rarely consistent and different jurisdictions have written their own legal and ethical guidelines. In recent years, a number of standards have been developed by health organisations wishing to collate and exchange medical data. While these rigorous standards are appropriate for handling sensitive patient details, they are unwieldy and unnecessarily limiting when it comes to recording broader consumer behaviours.

For the vast majority of people, a middle way is required: a standard that protects their right to privacy while allowing their data to be collected and interpreted in ways they have signed up for. The Coelition standard offers such a solution, paving the way for valuable services for millions of consumers.

Its creation is timely. The internet and associated mobile communications are now a mature set of technologies on which many services are being built. Although early internet pioneers based their business models on a set of initially disruptive innovations, they are now mature corporations with a high degree of operational stability. Accordingly, many have an entrenched attitude about their right to operate as they wish, in the face of growing concerns about privacy expressed by individual citizens and governments. The use of highly targeted advertising based on a detailed analysis of an individual's surfing or shopping habits will not be surrendered lightly – there is too much revenue at stake.

More light can be shed on this issue by returning to the reductionist versus the expansionist view of personal information, held by US and EU regulators respectively. For most US-based companies, the protective position of European jurisdictions seems unnecessary and limiting. But for most European businesses it would be unthinkable to set up a business model that did not, from the outset, include the strictest levels of personal information protection. The Coelition standard is a new and conscious attempt to reconcile these polar opposites. It is designed to allow brands to get up close and personal with their consumers, wherever they live, while guaranteeing the highest levels of privacy.

The architects of Coelition bring together a diverse set of skills and work in a wide cross-section of industry segments. To create a new way of understanding consumers in the digital age, it is not enough to have access to superb software engineering and hardware device expertise. It is also vital to underpin the whole endeavour with a deep understanding of the habits, needs, rituals and attitudes of ordinary consumers.

Our work takes full account of the pressing need to tackle today's social issues: obesity, diabetes, and shortages of energy, food and water. In that sense Coelition is very much of its time: it has arisen from a confluence of global challenge, digital technology opportunity and consumer insight. Yet at the same time it is timeless, built on an acknowledgement of the atomistic nature of the everyday lives of all human beings.

Today it is difficult to avoid articles and adverts that mention big data. Although there is no doubt the issue is here to stay, much of the debate is rudimentary. Coelition is not about big data. The standard makes a very conscious distinction between *data* and *information* as they are two different things and need to be treated as such. This distinction is rarely made in mainstream debate, but it is essential for clarity.

Coelition allows organisations and individuals to use the technologies that drive big data – pervasive sensors, mobile computing devices and cloud servers – without lumping everything indiscriminately together. Instead, the data is transformed into atoms, each and every one being a self-contained and unique piece of information. An atom already has meaning in a way that an isolated piece of data does not. It is inherently secure, so that even if it is lost, copied or diverted, no privacy is infringed. And yet when these atoms are collated, they can generate powerful insights and services.

Coelition will enable many things, but perhaps the biggest impact will be on the creation of insights and positive behaviour change – benefiting organisations and consumers alike. For commercial organisations, use of the standard will help inform the next generation of marketing, while for non-profit and public health organisations it will aid effective engagement with real individuals. For consumers, the standard will provide a new way to understand and change their own lives, without fear of Big Brother.

Staying open

Coelition is an open standard. To understand what this means in practice, it is useful to consider how standards are arrived at. The global marketplace has led to the development of thousands of international standards, describing key elements of the safety, usability or environmental impact of millions of products. These standards are usually designed by international committees representing leading industrial nations. Over the past half-century, many previously national standards have been adopted, usually with some minor modifications, as international standards administered by the International Organization for Standardization (ISO).

For consumers, ISO standards ensure that products and services are safe and of good quality. For businesses, they provide a foundation for uniformity, interoperability and fair access to global markets. However, the process of creating an ISO standard is very slow, requiring consensus between hundreds of different countries and organisations. It is not surprising that other, much quicker, ways to create standards have sprung up – a key example being the development of protocols for the internet during the past twenty-five years. Often these are forged by a small group of individuals who see both a specific issue to be solved and an elegant technical solution. Rather than trying to persuade the ISO to adopt these standards, they choose another route – they make the standards *open*.

Open standards can provide multiple stakeholders (including large corporations, governments and SMEs) with the conceptual framework and technical means necessary for integration and collaboration. If set up properly, open standards can help prevent a single self-interested party from seizing control. They can facilitate competition by lowering the cost of entry to a marketplace, as well as stimulating innovation by companies wishing to engage with the ecosystem created by the standard while also differentiating themselves. Meanwhile end users can benefit from the resulting interoperability and freedom from reliance on a single supplier.

The term 'open' can mean different things to different people, and not all open standards are equally open. Coelition is an example of an open standard that embraces the following principles:

- Its conceptual foundation does not represent the view-point of a single entity.
- Its progress and development is not controlled by a single entity.
- It is available for use by anyone who signs up to its aims, under reasonable and non-discriminatory legal and commercial terms.

Although Coelition is primarily an industry standard, it has also been designed to evolve into a consumer-visible brand. The name 'Coelition', its logo and its communication style are intended to be a user-friendly alternative to relatively impersonal and technical protocols such as the IEEE standards.

Coelition is both the name of the standard and of the organisation that publishes and administers it. It is a company limited by guarantee with articles of association that limit the redistribution of funds and ensure it acts transparently.

Coelition

How does the standard operate?

Coelition is a membership organisation. Its members decide how to disseminate the expertise and information incorporated in the standard. There are a range of Coelition membership levels, each with a set of conditions and obligations as well as differing levels of confidentiality. Membership agreements contain clauses relating to:

1 *Confidentiality and intellectual property.* Members help define the scope of intellectual property rights (IPR) within the consortium. The primary aim is to ensure that members agree not to take legal action against each other regarding any IP that has been created by or transferred into the consortium. A key element is an agreement to keep the core features of the standard, such as the details of the COEL, confidential.
2 *Cross-licensing.* Members sign up to the principle of allowing a fair, reasonable and non-discriminatory zero-royalty licence to other members for IPR they hold that is within the scope of Coelition. This does not preclude individual companies protecting their own commercial interests, but it does encourage interoperability and controlled collaboration.
3 *Copyright grants.* Members who are active in creating and modifying the standard will devise associated new text and images, the copyright for which will be transferred to Coelition.

To maintain and enhance the value of the standard, Coelition will robustly protect its trademarked assets, and pursue all copyright infringements and breaches of confidentiality.

Membership classes

There are three classes of Coelition member: service providers, developers and data engines. Each has a distinct scope and set of obligations.

The vast majority of organisations registering as members will be *service providers* of one sort or another. They will provide services directly to consumers or to commercial and research organisations that depend on information about individuals. In all cases, a Coelition-compliant service provider has identity protection obligations to the individual. There will be a wide range of different types of service provider, including:

- Brands wishing to provide trusted, personalised services to their customers.
- Commercial and non-profit behaviour change organisations offering services in health and wellbeing.
- Commercial research and market research organisations that use Coelition as the base platform for their services, with the addition of other data sources such as video records.
- Organisations such as universities and public health groups with an interest in epidemiology and research into behaviour recording and change.

Coelition

The publication and adoption of the Coelition standard will prompt the creation of a wider ecosystem of third-party device and app creators. These companies will be able to register with Coelition as *developers*, giving them access to free resources such as software development kits and support from Coelition-registered data engines (see below). Once a developer has created a Coelition-compliant app, it will submit it to Coelition to be tested for interoperability and compliance with the standard.

A *data engine* is a B2B organisation providing data-handling services to other businesses. It has no relationship with individual consumers. Any Coelition-compliant data engine is obliged to register with the consortium and maintain a Coelition-compliant system. This includes the following obligations:

1 To maintain always-on single entry points for atoms.
2 To receive these atoms free of charge.
3 To host a device registration service.
4 To provide limited free services for registered app and hardware developers.
5 To provide reduced-price services for legitimate non-profit organisations.
6 Not to hold any user's personal information.
7 Not to act as a service provider itself.

In the early development phase of the standard there will be only one data engine. After full publication of the standard we expect more to register and to develop their own standard-compliant offerings. In principle, we can also anticipate the development of a controlled market-place for the Coelition atom databases created by data engines, as they never contain directly identifying personal information. The way Coelition manages personal information was discussed in Chapter 7. To illustrate how this identity protection will work in practice, Figure 10.1 shows the sequence of events for a consumer signing up to a Coelition-based service. The sequence is as follows:

(a) The consumer signs up to a service provider and shares with it personal information such as their name and address.
(b) The service provider requests and receives a unique digital identifier for that consumer from the Coelition IDA.
(c) The consumer registers a device, such as a mobile phone, with the service provider.
(d) The service provider links the device with the consumer's digital identifier and registers this with the data engine.

The consumer can now use their device to capture information about their behaviour. This travels directly to the data engine as atoms. Note that these atoms do not contain any personal information and therefore the data

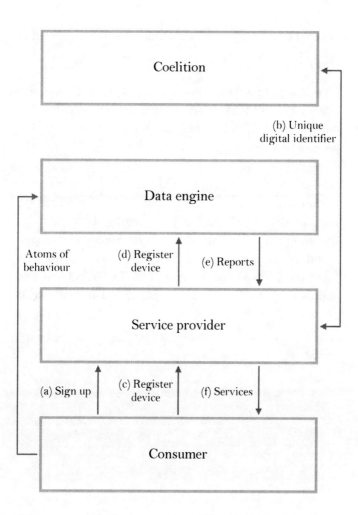

Figure 10.1 The basic operation of the Coelition ecosystem.

engine never knows which individual the digital identifier refers to.

(e) The data engine creates reports, based on the consumer's behaviour, and sends them to the service provider.

(f) The service provider uses these reports to provide a personalised service to the consumer.

Coelition has been designed to be as widely used as some of the core open standards on which the internet and web are built. It creates a new way of thinking about everyday life, and provides a humanising layer for our computers, mobile phones and other devices so they work seamlessly together. It gives people access to valuable, reliable and actionable behavioural data at unparalleled depth, scale and pace.

Coelition

Manifesto 11

THE Coelition standard describes a holistic way to measure and record human behaviour. It encompasses every activity inherent in daily living.

The standard helps businesses and other organisations to devise new services, and wider society to explore the future direction of digital behaviour capture. It provides a benchmark against which individuals can evaluate and choose trusted suppliers of personalised services.

The following principles are each beneficial in their own right, but taken as a whole they lead directly and uniquely to the Coelition standard.

1 **Daily life is granular.** Our daily lives are made up of a finite number of behaviours which have a natural granularity.

2 **Classify before quantifying.** It is both possible and desirable to classify, name and code all the daily behaviours that make up an individual's life.

3 **Measure only observable behaviours.** Measuring and recording only the behaviours that are observable provides the most robust and accurate view.

4 **Distributed intelligence.** Devices are the interfaces between humans and digital systems. We should use their intelligence to discriminate behaviours locally, to give modularity and rich information.

5 **Don't waste bandwidth.** Use the most economical means available to communicate the right information.

6 **Humanise technology.** Humanise technology and knowledge rather than digitise life.

7 **Separate personal and behavioural information.** Real information about behaviours should never be sent or unnecessarily stored with the personal information that directly identifies an individual.

8 **No free trade of behavioural patterns.** A database of recorded behaviours will contain patterns capable of identifying individuals. Any market for such information must be carefully controlled.

9 **Use open standards to drive transparency and innovation.** Standards are necessary for transparency, interoperability and depth of information. Making these standards open promotes innovation.

10 **Keep data collection and feedback tasks distinct.** The relaying of information and insights to an individual should be in the right form, context and medium. It does not need to use the device that made the measurements.

11 **Change behaviour one step at a time.** Behaviour change is highly complex, and even simple interventions can yield many different outcomes. Approaches need to be tested within a system that can assess and communicate the impact of small steps.

12 **Trusted services are provided by strong brands.** When dealing with information, service always comes before a physical product. Trusted brands provide the best route to meaningful dialogue in a mass consumer market.

Notes

The following notes connect a phrase in the main book text to a reference, usually a book, scientific paper or website, that is either supporting evidence for what we have said or we think is important for anyone wanting to follow a line of enquiry. The notes provide some context to the reference or additional material that would have interrupted the flow of the main text.

Chapter 1 - Daily Life

WE LIVE OUR LIVES IN DIGITAL NETWORKS
Sandy Pentland has written extensively in the area of big data and its impact on society. In this paper, with three other authors, he uses the term 'digital breadcrumbs' to describe the imprints we leave on the world and how this can be used to determine information about our behaviours.

Alex (Sandy) Pentland, David Lazer, Devon Brewer and Tracy Heibeck, *Using Reality Mining to Improve Public Health and Medicine* (Princeton: Robert Wood Johnson Foundation Whitepaper, 2009).

MANY PEOPLE SEE THIS AS A HUGE OPPORTUNITY
Gordon Bell of Microsoft Research collected all of his digital memories online between 1998 and 2007 in a project called Mylifebits. He and Jim Gemmell went on to write a book about the experience.

Gordon Bell and Jim Gemmell, *Your Life, Uploaded,* originally published as *Total Recall* (New York: Plume, 2010).

THE QUANTIFIED SELF MOVEMENT

The Quantified Self movement brings together enthusiasts who use devices to measure, record and share data about their personal physical and mental states. Also known as Personal Informatics and Self-tracking.

<http://quantifiedself.com/>

EXPERIMENTED WITH A WRIST-WORN SENSOR

There is a long history of productive and sometimes dangerous self-experimentation in science. Lawrence Altman's book is a good starting point to understand more about this.

Lawrence K. Altman, *Who Goes First? The Story of Self-Experimentation in Medicine*, new edn. (Berkeley: University of California Press, 1998).

The following paper by Jack Michael gives an articulate critique of statistical methods employed in these and similar cases.

Jack Michael, 'Statistical Inference for Individual Organism Research: Mixed Blessing or Curse?', *Journal of Applied Behavior Analysis*, Vol. 7, No. 4, (1974) pp. 647-653.

The wrist-worn sensor that Joss has been wearing continuously since August 2012 is a GENEActiv. Originally created by Unilever R&D and now a product of Activinsights Ltd, this device records acceleration, light and temperature at 100Hz. It is used in a wide range of research applications across the world.

<http://www.geneactiv.co.uk>

BIG DATA IS INCREASINGLY ABOUT REAL BEHAVIOUR

Sandy Pentland talks about his view of big data in this interview on John Brockman's website Edge.

<http://www.edge.org/conversation/reinventing-society-in-the-wake-of-big-data>

TWO IN THREE ADULTS IN THE US

The Pew Research Center uses public opinion polls and social science research to understand attitudes and trends shaping America and the world. Their 'Tracking for Health' report is based on a telephone survey of 3,014 adults living in the US.

Susannah Fox and Maeve Duggan, *Tracking for Health* (Washington, DC: Pew Research Center, 2013).

A MIXTURE OF REPUGNANCE AND DISBELIEF

This publication from Unilever sets out its practical view on achieving meaningful behaviour change. In the introduction Paul Polman states: 'Successful change comes from a real understanding of people, their habits and their motivations.'

Unilever, *Inspiring Sustainable Living, Expert Insight into Consumer Behaviour and Unilever's Five Levers for Change*, (London: Unilever, 2011).

IF ONLY WE CAN ARRIVE AT A COMMON STANDARD

An example of some good thinking on standards for personal informatics can be found in the following paper by Ian Li and others.

Ian Li, Anind Dey, and Jodi Forlizzi, 'A Stage-based Model of Personal Informatics Systems', *CHI 2010 Proceedings of the SIGCHI Conference on Human Factors in Computing Systems*, (2010) pp. 557-566.

Chapter 2 - Atoms

LET US RECORD THE ATOMS AS THEY FALL

Virginia Woolf (1882-1941) was one of the leading modernist writers of the twentieth century and an influential intellectual in London literary circles.

Virginia Woolf, 'Modern Fiction' in *The Essays of Virginia Woolf. Volume 4: 1925 to 1928*, ed. by Andrew McNeille (London: The Hogarth Press, 1984).

WOOLF HERSELF, AND HER CONTEMPORARIES

Liesl Olson explores the way modernist writers James Joyce, Virginia Woolf, Gertrude Stein and Wallace Stevens deal with everyday life against the backdrop of war and upheaval.

Liesl Olson, *Modernism and the Ordinary*, (New York: Oxford University Press, 2009).

Bryony Randal also looks at the writings of Woolf and Stein, as well as Dorothy Richardson, to show how these writers see daily time and everyday life as rich subjects in and of themselves.

Bryony Randall, *Modernism, Daily Time and Everyday Life*, (Cambridge, UK: Cambridge University Press, 2011).

ATOMIC HYPOTHESIS
John Dalton (1766-1844) was an English polymath chemist, physicist and meteorologist who is remembered both for his work on colour blindness and as the father of modern atomic theory.

Henry Guerlac, 'Quantification in Chemistry', in *Quantification: a History of the Meaning of Measurement in the Natural and Social Sciences*, ed. by Henry Woolf (Indianapolis: Bobbs-Merrill, 1961).

IF, IN SOME CATACLYSM
Richard Feynman (1918-1988) was an American theoretical physicist. Although his technical work is difficult for many lay people to understand he was a committed educator and populariser of the power of science.

Richard Feynman, Ralph Leighton and Matthew Sands, *The Feynman Lectures on Physics* (Reading, MA: Addison-Wesley, 1964), pp. 1-2.

ALL THE DIVERSE FORMS OF LIFE ON EARTH
American biologist E.O. Wilson makes the following point about the relationship between the diversity of life and the laws of physics: 'The study of living things at the molecular level established what may fairly be called the *First Law of Biology*, that all the entities and processes of life are obedient to the laws of physics and chemistry... the *Second Law of Biology*, that all entities and processes of life were created by evolution through natural selection.'

Michael R. Canfield, Edward O. Wilson, George B. Schaller, Bernd Heinrich and Bernd Kaufman, *Field Notes on Science and Nature* (Cambridge, MA: Harvard University Press, 2011).

USING A ROUGH ORDER-OF-MAGNITUDE ESTIMATE
In problems that have many unknown quantities it is useful to get an 'order-of-magnitude' estimate of the answer. This type of estimation ignores some of the details of the quantities if a problem works with powers of ten: 10, 100, 1000, etc. For tutorial examples and useful tricks see:

Sanjoy Mahajan, S*treet-Fighting Mathematics: The Art of Educated Guessing and Opportunistic Problem Solving* (Cambridge, MA: MIT Press, 2010).

Chapter 3 - What

BEFORE WE CAN QUANTIFY SOMETHING

Taxonomy is no longer considered cutting edge, but it remains a noble foundation for much of biological science. For an entertaining history and discussion of biological taxonomy and some of the issues it raises, see:

> Carol Kaesuk Yoon, *Naming Nature: The Clash Between Instinct and Science* (New York: W.W. Norton & Company, 2009).

THIS MEANS THAT IN PRINCIPLE

Currently about 1.2 million species are catalogued, based on the cumulation of 250 years of scientific endeavour. It is difficult to estimate how many species remain unknown; the following paper predicts that there are probably 8.7 million Eukaryotic (i.e. multi-cellular) species on Earth.

> Camilo Mora, Derek Tittensor, Sina Adl, Alastair Simpson and Boris Worm, 'How Many Species Are There on Earth and in the Ocean?', *PLOS Biology*, Vol. 9, No. 8, (2011).

CARL LINNAEUS

Carl Linnaeus (1707-1788) published *Species Plantarum* (The Species of Plants) in 1753. It is widely credited as the starting point for the plant nomenclature we still use today. Recent developments of the Linnean system by the Angisperm Phylogeny Group of scientists have incorporated genetic characteristics as well as the more traditional plant anatomy.

> Angiosperm Phylogeny Group III, 'An update of the Angiosperm Phylogeny Group classification for the orders and families of flowering plants: APG III', *Botanical Journal of the Linnean Society*, Vol. 161, (2009) pp. 105-121.

THE COEL IS A NEW EVENT-BASED TAXONOMY

A number of other classification schemes have been created over the years that refer to aspects of everyday living. Lawton and Brody have developed a listing of around 30 Activities of Daily Living (ADLs) that are used to assess the lives of older people.

> M. Powell Lawton and Elaine M. Brody, 'Assessment of Older People: Self-Maintaining and Instrumental Activities of Daily Living', *Gerontologist*, Vol. 9, (1969) pp. 179-186.

Meanwhile Ainsworth *et al.* have created a compendium of many different activities and an estimate of the physical energy levels required for each.

Barbara Ainsworth, William Haskell, Melicia Whitt, Melinda Irwin, Ann Swartz, Scott Strath, William O'Brien, David Bassett Jr, Kathryn Schmitz, Patricia Emplaincourt, David Jacobs Jr and Arthur Leon, 'Compendium of Physical Activities: an Update of Activity Codes and MET Intensities', *Medicine and Science in Sports and Exercise*, Suppl., Vol. 32, No. 9, (2000) pp. 498-504.

ENSURING THE LIST IS BOTH MUTUALLY EXCLUSIVE
The MECE approach is a core skill in management consulting. The use of MECE in consulting appears to originate in the work of Barbara Minto with McKinsey and other consulting firms. It remains an integral part of the approach taught to McKinsey recruits.

Barbara Minto, *The Pyramid Principle: Logic in Writing and Thinking*, 3rd ed. (London: Prentice Hall, 2008).

Ethan Rasiel explains that: 'MECE (pronounced "me-see") stands for "mutually exclusive, collectively exhaustive" and it is a sine qua non of the problem-solving process at McKinsey. MECE gets pounded into every new associate's head from the moment of entering the firm.'

Ethan Rasiel, *The McKinsey Way* (McGraw-Hill, 1999), p. 6.

The core elements of the MECE concept can be traced back to the theologian Blessed John Dun Scotus (1266-1308) who notes in his *De Primo Principio* (A treatise on God as first principle): 'For a division to be clear it is necessary (1) that the members resulting from the division be indicated and thus be shown to be contained in what is divided, (2) that the mutually exclusive character of the parts be manifest, and (3) that the classification exhaust the subject matter to be divided.'

<http://www.franciscan-archive.org/scotus>

THE ISOTYPE APPROACH
Isotype (International System of Typographic Picture Education) was a complete pictorial language and approach to pictorial statistics, originally developed in Austria between 1925 and 1934 by Otto Neurath and his co-workers, including the artist Gerd Arntz. In the Isotype system, each of the pictorial symbols represents a number. To indicate a higher number, more symbols of exactly the same size are

used. Meanwhile, colour is used to discriminate between symbols of the same basic form.

Otto Neurath, *From Hieroglyphics to Isotype: A Visual Autobiography* (London: Hyphen Press, 2010).

One of the key elements in the Isotype approach was the role of people, called *transformers*, who were able to map raw numerical and statistical data into a pictorial representation. Isotype remains deeply interesting to typographers and students of the history of scientific graphics.

Robin Kinross and Marie Neurath, *The Transformer: Principles of Making Isotype Charts* (London: Hyphen Press, 2009).

A SET OF EMOTIONALLY ENGAGING ITEMS

Strictly speaking, the COEL icons are better thought of as *ideograms*. Unlike an icon, which usually symbolises a single object, an ideogram can illustrate an idea, an action, an activity or an emotion. The icons contain only what is essential to communicate their core meaning or message. The creation of our initial set of COEL icons was based on observational drawing of the human form, the use of perspective and isometric planes, and the use of line and tone to represent form. *The Noun Project* is a good example of an icon catalogue with a strong design ethic which was founded by Edward Boatman and Sofya Polyakov. In contrast, Kreston Bjerg's *Phenomenalog* focuses on how glyphs can have application in daily life with a much clearer bias towards action.

<http://thenounproject.com>
<http://www.phenomenalog.dk>

Chapter 4 - *When*

EVENTS ARE PERCEIVABLE BUT TIME IS NOT

The quote is from the title of James J. Gibson's 1975 paper. Gibson also developed the concept of 'optical flow', which describes visual stimulation in animals and posits that behaviour (such as flying speed in bees) is tuned to maximise the use of their visual capacities.

James J. Gibson, 'Events are Perceivable but Time is not', in *The Study of Time II* ed. by J.T. Fraser and N. Lawrence (New York: Springer-Verlag, 1975), pp. 295-301.

PHYSICAL AND PERSONAL TIME
In anthropology texts it is possible to find a bewildering number of different types of time. A good example is found in the book below.
Edward T. Hall, *The Dance of Life*, 2nd edn. (New York: Anchor Books, 1989).

CLOCKS MEASURE PHYSICAL TIME
We deliberately do not include a discussion of Einstein, relativity or space-time as Newtonian physics is a sufficient approximation for our purposes. For a readable introduction to these ideas see:
Russell Stannard, *Relativity: A Very Short Introduction* (Oxford: Oxford University Press, 2008).

THE SECOND IS THE DURATION OF 9,192,631,770 PERIODS
A second is an arbitrary, but very precisely defined, period of time that has been internationally agreed as part of the SI system of units.
Bureau International des Poids et Mesures, *The International System of Units (SI)*, 8th edn. (Sèvres: Le Bureau, 2006).

PERSONAL TIME IS DEPENDENT ON OUR STATE OF MIND
There are many examples in experimental psychology of how we perceive the passing of time under different psychological and environmental conditions. Paul Fraisse's book provides a thorough account of work up to the 1960s and Robert Ornstein describes a number of intriguing experiments:
Paul Fraisse, *The Psychology of Time*, trans. by Jennifer Leith (London: Eyre & Spottiswode, 1964).
Robert E. Ornstein, *On the Experience of Time* (London: Penguin, 1969).

TIME-RECKONING ACCURACY
Most authors agree that we have a temporal horizon of a small number of seconds that constitute our perception of the 'now' and that we react to stimuli in a few hundred milliseconds. The Coelition standard allows the time stamp and duration of an event to be recorded to an accuracy of one second. As well as Fraisse and Ornstein above, see the more recent work by Southerton.
Dale Southerton, 'Analysing the Temporal Organization of Daily Life: Social Constraints, Practices and their Allocation', *Sociology*, Vol. 40, No. 3, (2006) pp. 435-454.

PERSONAL TIME PLAYS A FUNDAMENTAL ROLE

Gell provides a useful account of different ways to think about time from a meta-physical perspective while grounding it in anthropology.

Alfred Gell, *The Anthropology of Time* (Oxford: Berg, 1992).

DIACHRONIC AND SYNCHRONIC

The terms diachronic and synchronic were introduced by the linguist Ferdinand de Saussure. A good example of their use by Lévi-Strauss can be found in the following paper. It is worth noting Gell's (see above) criticism of any claims that time can be seen, in reality, as anything other than a continuous linear progression.

Claude Lévi-Strauss, 'The Structural Study of Myth', *The Journal of American Folklore*, Vol. 68, No. 270, (1955) pp. 428-444.

ADULTS SLEEP FOR APPROXIMATELY EIGHT HOURS

For two excellent introductions to the study of sleep see:

Till Roenneberg, *Internal Time* (Cambridge, MA: Harvard University Press, 2012).

Steven W. Lockley and Russell G. Foster, *Sleep: A Very Short Introduction* (New York: Oxford University Press, 2012).

CHRONOTYPE

The Munich ChronoType Questionnaire is used to assess chronotype in a questionnaire format. The URL below is supported by Till Roenneberg and will return an assessment of your chronotype by email. Joss's results from this questionnaire agreed exactly with the analysis made from wrist temperature.

<http://www.thewep.org>

Chapter 5 - How

THE WEBSITE HUNCH.COM

Hunch.com provides personalisation services on the internet, linking different types of preferences. Its stated ambition is to create a 'Taste Graph' of the entire web.

<http://blog.hunch.com/?p=47384>

BALANCE OF IMAGES, INSIGHT, SPEED AND PLAYFULNESS

The use of gameplay theory and design is a growing topic in behaviour change, for example see:

Sebastian Deterding, Rilla Khaled, Lennart Nacke and Dan Dixon, 'Gamification: Toward a Definition', *Proceedings of the 2011 Annual Conference Extended Abstracts on Human Factors in Computing Systems*, (2011).

OUR RECOLLECTION OF EVENTS IS OFTEN SKETCHY

The Day Reconstruction Method (DRM) provides a framework for building a picture of activities and emotions from a subject's previous day.

Daniel Kahneman, Alan B. Krueger, David A. Schkade, Norbert Schwarz and Arthur A. Stone, 'A Survey Method for Characterizing Daily Life Experience: The Day Reconstruction Method', *Science*, Vol. 306, (2004) pp. 1776-1780.

Daniel Kahneman's book below puts the DRM paper in context with his other work and is a worthwhile read.

Daniel Kahneman, *Thinking, Fast and Slow* (London: Allen Lane, 2011).

The Multimedia Activity Recall for Children and Adolescents approach described in the following paper provides another approach to reconstructing a previous day's activity, with a focus on physical activity.

Kate Ridley, Tim S. Olds and Alison Hill, 'The Multimedia Activity Recall for Children and Adolescents (MARCA): Development and Evaluation', *International Journal of Behavioral Nutrition and Physical Activity*, Vol. 3, No. 10, (2006).

ALLOW OBSERVED HUMAN EVENTS TO BE CODED

There are a number of systems available for the manual coding of recorded video to assist with classification and analysis. An example from Mangold is given below.

<http://www.mangold-international.com/software/interact/>

ETHNOGRAPHIC RESEARCH

An ethnography is a representation of the culture of a people in written or graphical form. The term has been extended by commercial researchers to cover the methodology of direct and close observation of subjects.

ATOM-RETURNING APPLICATIONS ENGAGE THE USER

The approach of interrupting a subject's day at regular intervals to ascertain their state of mind was initially developed in the 1980s and is described in the following paper. Mobile phones now provide the ideal platform to run studies using these methods.

Mihaly Csikszentmihalyi and Reed Larson, 'The Experience Sampling Method', in *Naturalistic Approaches to Studying Social Interaction. New Directions for Methodology of Social and Behavioral Science*, ed. by H.T. Reis, Vol. 15, (San Francisco: Jossey-Bass, 1983), pp. 41-56.

THIS BIG DATA

In spite of the current hype around 'big data', from a technical point of view it is simply the use of well-known statistical techniques for finding correlations or patterns in data but applied to enormous unstructured data sets. What has generated the hype is the sheer scale of the data and the fact that the data is being created mainly by individual consumers. The MIT Technology Review estimates that 75% of all digital data is now created by consumers, not big scientific experiments or financial institutions.

<http://www.technologyreview.com/news/514351/has-big-data-made-anonymity-impossible/>

MEASUREMENT IS A DEFINED PROCESS

A clear explanation of the role of measurement and quantification in the scientific process is given by Samuel Wilks. According to Wilks a measurement process should meet the following three desiderata:

1. Measurement has to be an operationally definable process of associating a real number, with appropriate units, to the aspect of the problem we are measuring. This means that a recipe can be written down that describes exactly what steps have to be taken to perform a measurement. This is a strict and very distinctive requirement. It implies that the measurement process is not itself a random or creative process. The order in which the steps are taken has huge significance.

2. A measurement process requires reproducibility of the outcome. If the measurement process is repeated in exactly the same manner and order on an object that has not changed since the last time it was measured we expect the same measured value. However, common experience tells us that often repeated measurements will not be identical and this will lead to a distribution of values. *(cont.)*

3. The third important aspect for a measurement process is that of validity or accuracy. How closely does the measured value come to the true value?

Samuel S. Wilks, 'Some Aspects of Quantification in Science', in *Quantification: a History of the Meaning of Measurement in the Natural and Social Sciences*, ed. by Henry Woolf (Indianapolis: Bobbs-Merrill, 1961).

HUMAN BEINGS HAVE BEEN USING STRUNG BEADS
Marian Vanhaeren and her colleagues have found shell beads with human fossils dating from 100,000 to 135,000 years ago. She has gone on to show how in beads 75,000 years old there is evidence of changing styles of wear.

Marian Vanhaeren, Francesco d'Errico, Chris Stringer, Sarah L. James, Jonathan A. Todd and Henk K. Mienis, 'Middle Palaeolithic Shell Beads in Israel and Algeria', *Science*, Vol. 312, (2006) pp. 1785-1788.

BRAD THE TOASTER
Brad was an extension of Simone Rebaudengo's graduation project while on an internship with Haque design.

<https://twitter.com/BradToaster>

ROUGHLY EQUIVALENT IN SIZE TO A SINGLE MOUSE CLICK
A USB mouse click using the HID protocol contains 64 bytes.

<http://www.usb.org/developers/devclass_docs/HID1_11.pdf>

NEW IMAGE-ANALYSIS APPROACHES CAN PROVIDE FOOD TYPE AND QUANTITY
In general, image analysis is computationally intensive and successful applications require careful method development to ensure that a theoretical approach can be made practically useful. The following paper is an interesting example:

Mingui Sun, Qiang Liu, Karel Schmidt, Lei Yang, Ning Yao, J.D. Fernstrom, M.H. Fernstrom, James P. DeLany and R.J. Sclabassi, 'Determination of Food Portion Size by Image Processing', *Engineering in Medicine and Biology Society*, (2008) pp. 871-874.

Chapter 6 - Why

FROM THESE ANALYSES

Although reductionism allows the whole process of pool ball movement to be explained and understood using physics, the complex reality of what happens when pool balls collide is not trivial. For comprehensive analysis as well as normal- and high-speed videos of pool table physics see the work of Dave Alciatore of Colorado State University.

<http://billiards.colostate.edu/>

REDUCTIONISM IS LESS USEFUL

The development of scientific concepts and approaches to the study of complex phenomena is relatively new. A good popular account of some of the pioneers of complexity or chaos theory is given by James Gleick:

James Gleick, *Chaos: Making a New Science*, (London: Vintage, 1997). For a more recent discussion about the limits of reductionism and determinism in biology, see:

Fulvio Mazzocchi, 'Complexity in Biology: Exceeding the Limits of Reductionism and Determinism Using Complexity Theory', *EMBO Report*, Vol. 9, No. 1, (2008) pp. 10-14.

THE LAW OF UNINTENDED CONSEQUENCES

This 'law' has been discussed by philosophers and economists for centuries, but it was brilliantly explained by the American sociologist Robert K. Merton (1910-2003) in the following paper. He promised to write a book about this law and worked on the subject until he died in 2003. A book based on the material has not been published.

Robert K. Merton, 'The Unanticipated Consequences of Purposive Social Action', *American Sociological Review*, Vol. 1, No. 6, (1936) pp. 894-904.

TWO AMERICAN SOCIAL SCIENTISTS

The three-part summary quoted here is found in the authors reminiscences of the publication of the book in *Current Contents Social & Behavioural Sciences*. The original book is:

Bernard Berelson and Gary A. Steiner, *Human Behavior: An Inventory of Scientific Findings*, (New York: Harcourt, Brace & World, 1964).

A TANGLED MESH OF CAUSES

Nobel laureate Clive Granger (1934-2009) developed a statistical method to determine whether one stream of data is useful in predicting another. This very specific definition of causality is useful in financial analysis but has limitations even when just looking at two data sources.

Clive W.J. Granger, 'Investigating Causal Relations by Econometric Models and Cross-spectral Methods', *Econometrica*, Vol. 37, No. 3, (1969) pp. 424-438.

IN RECENT YEARS THERE HAS BEEN A RESURGENCE

An example from the recent resurgence in observational natural history is given below.

Rafe Sagarin and Anibal Pauchard, *Observation and Ecology: Broadening the Scope of Science to Understand a Complex World*, (Washington: Island Press, 2012).

MARINE BIOLOGIST ED RICKETTS

Ed Ricketts (1897-1948) was described by his sister as having 'a mind like a dictionary'. He was a friend of John Steinbeck, providing the model for 'Doc' in his 1954 novel *Cannery Row*. Steinbeck wrote: 'He wears a beard and his face is half Christ and half satyr and his face tells the truth'. Ricketts died prematurely in 1948 after his car was hit by a train in Monterey.

Eric Enno Tamm, *Beyond the Outer Shores: The Untold Odyssey of Ed Ricketts, the Pioneering Ecologist Who Inspired John Steinbeck and Joseph Campbell*, (New York: Four Walls Eight Windows, 2004).

LEADING US TO OVER-FIT THE DATA

Gerd Gigerenzer provides an excellent account of how computer-based statistical analysis methods can place too much weight on the information they have, compromising their predictive abilities.

Gerd Gigerenzer and Henry Brighton, 'Homo Heuristicus: Why Biased Minds Make Better Inferences', *Topics in Cognitive Science*, No. 1, (2009) pp. 107-143.

COMPARED TO WHAT

This formulation of the key question to be answered in quantitative reasoning is from Edward Tufte.

Edward Tufte, *Envisioning Information*, (Cheshire, CT: Graphics Press, 1990), p. 67.

ALTHOUGH THIS APPROACH MAY SEEM CONTRIVED

The analogy drawn between the problems of medical diagnosis and deductive and inductive inferences is taken from this paper by Steven Goodman.

Steven N. Goodman, 'Toward Evidence-Based Medical Statistics. 1: The P Value Fallacy', *Annals Internal Medicine*, Vol. 30, (1999) pp. 995-1004.

NEVER INVERT THE DATA

The iterative data analysis strategy, which we describe as 'never invert the data', is based on experience with a number of data inversion techniques used in imaging. In these applications, data inversion fails if there are a significant number of missing values or the data is 'noisy'. Examples of the iterative approach are given in:

Brian Buck and Vincent A. Macaulay, *Maximum Entropy in Action: A Collection of Expository Essays* (New York: Oxford University Press, 1991).

MAKE BEHAVIOURAL CHANGES

There are many approaches to behaviour change, most of which deal with changing a single behaviour at a time. The following references show some of this breadth.

Richard H. Thaler and Cass R. Sunstein, *Nudge: Improving Decisions About Health, Wealth and Happiness*, (London: Penguin, 2009).

Institute for Government and the Cabinet Office, MINDSPACE, Influencing Behaviour Through Public Policy (2010). <http://www.instituteforgovernment.org.uk/sites/default/files/publications/MINDSPACE.pdf>

B.J. Fogg, *Persuasive Technology: Using Computers to Change What We Think and Do*, (San Francisco: Morgan Kaufmann Publishers, 2003).

THE KEY ETHICAL MAXIM

This well-known maxim does not form part of the Hippocratic oath. Its specific Latin form *Primum Non Nocere* apparently first appeared in the 1860s and has been transmitted primarily by oral tradition since.

Cedric M. Smith, 'Origin and Uses of *Primum Non Nocere* — Above All, Do No Harm!', *The Journal of Clinical Pharmacology*, Vol. 45, No. 4, (2005) pp. 371-377.

WITH THE SIMPLEST OF VISUALISATIONS

Often referred to as Infovis, visualisation of information is a domain that brings together scientists, statisticians, computer programmers, designers and artists.

Zachary Pousman, John T. Stasko and Michael Mateas, 'Casual Information Visualization: Depictions of Data in Everyday Life', *IEEE Transactions on Visualization and Computer Graphics*, Vol. 13, Issue 6, (2007) pp. 1145-1152.

MAKE SENSE OF EVERYDAY LIFE

For a good introduction to the study of everyday life from a sociology perspective see Susie Scott's book.

Susie Scott, *Making Sense of Everyday Life*, (Cambridge, UK: Polity Press, 2009).

CREATE DETAILED AND AUTHENTIC NARRATIVES

The analysis of complex systems has thrown up a number of interesting new computational approaches over the past few decades. Many of the approaches used by a new breed of computational social scientists rely on agent-based computational models. In these approaches a number of autonomous *agents* are encoded in computer programs and their interactions followed. What is surprising is how a small set of simple rules of interaction between agents gives rise to complex and dynamic system behaviour that mimics the observed complexity and dynamics of real social systems.

John H. Miller and Scott E. Page, *Complex Adaptive Systems: An Introduction to Computational Models of Social Life* (Princeton & Oxford: Princeton University Press, 2007).

Chapter 7 - Who

DATA ENGINE DOES NOT HOLD PERSONAL INFORMATION

The preservation of the anonymity within the data engine is a very important goal for the overall system. This is controlled in much the same way as a nightclub: there is only one way in, there is a strong door policy and both are guarded.

TO DISCRIMINATE IS TO IDENTIFY

Below are just two examples of the many examples now available to prove this point.

Arvind Narayanan and Vitaly Shmatikov, 'Privacy and Security: Myths and Fallacies of "Personally Identifiable Information"', *Communications of the ACM*, Vol. 53, No. 6, (2010).

Lukasz Olejnik, Claude Castelluccia and Artur Janc, 'Why Johnny Can't Browse in Peace: On the Uniqueness of Web Browsing History Patterns', *5th Workshop on Hot Topics in Privacy Enhancing Technologies* (2012).

INDIVIDUALS HAVE BEEN RE-IDENTIFIED
Also see Arvind Narayanan and Vitaly Shmatikov above.

Paul Ohm, 'Broken Promises of Privacy: Responding to the Surprising Failure of Anonymization', *UCLA Law Review*, Vol. 57, (2010) p. 1701.

INFORMATION SETS ARE CRITICAL FOR SOCIETY
Sandy Pentland articulates this point very well.

Alex Pentland, 'Reality Mining of Mobile Communications: Toward A New Deal On Data', *The Global Information Technology Report 2008-2009: Mobility in a Networked World*, ed. by Soumitra Dutta and Irene Mia (Geneva: World Economic Forum, 2009) pp. 75-80.

DATA ENGINE'S RIGHT TO LINK, DATA MINE OR TRADE
The data engine has the right to link data sets that are required for the delivery of the Coelition standard but no others. It must not mine the data set in an attempt to synthesise personal information and it can only trade atoms with other data engines that are compliant with and accredited by Coelition.

CHANGING SOCIAL NORMS
It is very likely that the shift from a base assumption that 'my life is largely private and I choose what to disclose' to the idea that 'my life is largely public and I choose what to retain' has occurred in many areas. In fact the title of Peter Schaar's book, *Das Ende der Privatsphäre* (*The End of Privacy*) would suggest that leading thinkers believe this to be the case. This will be abhorrent to many conservative observers. Other readers may believe this is exactly the type of change required to solve environmental issues where the link between our actions and shared resources (air, water, energy) needs to be foremost in our minds. If this is a direction of social change then it will be anthropologists rather than technologists who are best placed to illuminate the road ahead.

Peter Schaar, *Das Ende der Privatsphäre* (Munich: Bertelsmann Verlag, 2007).

CONSTRUCTS THAT DESCRIBE PERSONAL INFORMATION
The original definitions in the arena of personally identifying information go back to 1890.

Samuel Warren and Louis Brandeis, 'The Right to Privacy', *Harvard Law Review*, Vol. 4, No. 5, (1890) pp. 193-220.

DATA PROTECTION APPROACHES OF THE US AND EUROPE
In the US the most stringent data protection legislation to date is in the health sector under the HIPAA regulations of 1996 which, among other items, specify eighteen personal health identifiers (PHIs) as a definition of personal data.

Health Insurance Portability and Accountability Act of 1996, 42 U.S. Supreme Court 1320d-9 (2010).

In the EU, the Data Protection Directive was adopted in 1995 and defines as personal data any information that can be used to identify an individual directly or indirectly.

Directive 95/46/EC of the European Parliament and of the Council of 24 October 1995 on the protection of individuals with regard to the processing of personal data and on the free movement of such data, L281, *Official Journal of the European Union* (1995).

In 2012, the White House administration of Barack Obama released a set of voluntary guidelines called the Consumer Privacy Bill of Rights and the EU unveiled draft legislation to supersede the existing directive.

'Consumer Data Privacy in a Networked World: A Framework for Protecting Privacy and Promoting Innovation in the Global Digital Economy', (Washington: The White House, 2012).

'Proposal for a Regulation of the European Parliament and of the Council of 25 January 2012 on the protection of individuals with regard to the processing of personal data and on the free movement of such data [General Data Protection Regulation] (2012)'.

For an overview of the differences in approach between the US and EU see Abraham L. Newman's book.

Abraham L. Newman, *Protectors of Privacy: Regulating Personal Data in the Global Economy* (Ithaca and London: Cornell University Press, 2008).

The UK Information Commissioner's Office also has a very good website on the practical implementation of data protection.

<http://www.ico.org.uk>

WHAT CONSTITUTES PERSONAL INFORMATION

The term deliberately excluded from the main text is Personally Identifiable Information (PII). It is heavily used by both technologists and the legal profession but is increasingly disputed. Also see Paul Ohm above.

Paul M. Schwartz and Daniel J. Solove, 'The PII Problem: Privacy and a New Concept of Personally Identifiable Information', *New York University Law Review*, Vol. 86, (2011) p. 1814.

REAL AND PERCEIVED RISKS

In December 2012 the photo sharing application Instagram updated its terms of service to clarify its rights over members' photos. The resulting consumer outcry was reported to have cost the company 25% of its membership and resulted in a 3% drop in the share price of parent company Facebook.

Chapter 8 - Where

ALTHOUGH HUMANS HAVE INHERITED FROM PRIMATES

Some of the most interesting insights into the human experience of 'lived-space' as a problem in philosophy and human behaviour were developed by the German philosopher Otto Friedrich Bollnow (1903-1991). His major work, *Human Space*, was published in German in 1963 and it has recently been translated into English.

Otto F. Bollnow, *Human Space*, (London: Hyphen Press, 2011).

As a young man Bollnow was a member of the Militant League for German Culture, a nationalistic anti-Semitic political society active during the Weimar Republic and the Nazi era. It is now rather difficult to judge Bollnow's political affiliations as he was clearly under the spell of the controversial German philosopher Martin Heidegger (1889-1976) at the time. Bollnow appears to have rewritten much of his work at a later date to remove the more overt Nazi elements.

BEFORE HUMANS HAD DEVELOPED MAPS

The songlines or dreaming tracks of indigenous Australians are not examples of analytical geography or maps, but they have allowed people to confidently travel significant distances across Australia for thousands of years. Each songline includes references to both sacred locations and physical locations to help guide the traveller (water holes, rocks, etc.).

<http://singing.indigenousknowledge.org/>

(cont.)

The British author Robert MacFarlane has written eloquently of his travels on the ancient network of long-distance paths in the UK and elsewhere.

Robert MacFarlane, *The Old Ways: A Journey on Foot*, (London: Hamish Hamilton, 2012).

OVER THE PAST FIVE HUNDRED YEARS

Maps are sophisticated conceptual tools. They allow the detailed extent of some part of the Earth's surface to be represented on a portable two-dimensional screen or piece of paper. Maps very often represent direction and distance, though some distort both distance and direction to aid the user, the London Tube map being a good example. Good-quality maps represent the Earth's surface with colours, shading, icons and text and the creation of great maps is both an art and a science. One of the classic texts on cartography that reveals some of the complexity of this craft is from Eduard Imhof.

Eduard Imhof, *Cartographic Relief Presentation*, (Redland, California: Esri Press, 2007).

A PLACE IS NOT SIMPLY A PARTICULAR LOCATION

The distinction between space and place and the description of the three component parts of place are from the excellent review article on the sociology of place by Thomas F. Gieryn.

Thomas F. Gieryn, 'A Space for Place in Sociology', *Annual Review of Sociology*, Vol. 26, (2000) pp. 463-496.

TEMPORARY OR PERMANENT MOVEMENT

This thorough review of the theoretical literature on migration covers politics, sociology and economics:

Jessica Hagen-Zanker, 'Why Do People Migrate? A Review of the Theoretical Literature', *Maastricht Graduate School of Governance Working Paper*, No. 2, (2008).

The gravity law of migration was first described by Harvard linguist and scholar George K. Zipf (1902-1950):

George K. Zipf, 'The $P1P2/D$ Hypothesis: On the Intercity Movement of Persons', *American Sociological Review*, Vol. 11, (1946) pp. 677-686.

Recently, advances in modelling human mobility patterns at different scales have been referenced and updated.

Filippo Simini, Marta C. Gonzalez, Amos Maritan and Albert-László Barabási, 'A Universal Model for Mobility and Migration', *Nature*, Vol. 484, No. 7392, (2010) pp. 96-100.

DATA MOBILITY PATTERNS OR MOTIFS

Human mobility can be described using a range of techniques including the motif approach with data from both surveys and mobile phone records:

Christian M. Schneider, Vitaly Belik, Thomas Couronné, Zbigniew Smoreda and Marta C. González, 'Unravelling Daily Human Mobility Motifs', *Journal of the Royal Society Interface*, Vol. 10, (2013).

Motifs are derived from a network analysis of the actual mobility patterns. To classify the networks, the places visited are treated as *nodes* in a network and each journey a directed *edge*, shown as an arrow on a network diagram. The network approach discards information about travel time between nodes, residence time at nodes, distance between nodes or the activity at the node. These networks are an abstracted minimal description of the individual's mobility pattern.

OBSERVED VISITATION DISTRIBUTION

This type of distribution of frequency versus rank is sometimes referred to as a 'Zipf' distribution:

George K. Zipf, *Human Behavior and the Principle of Least Effort*, (Cambridge, Massachusetts: Addison-Wesley, 1949).

Zipf originally identified the curve in his analysis of word frequencies, where he identified that the frequency of any word is inversely proportional to its rank in the frequency table. The same relationship seems to occur for other situations where a frequency is plotted versus rank (e.g. the population ranks of cities in different countries, the sizes of corporations, number of people with a given income versus ranking, etc.). In the case of the Zipf visitation distribution, the frequency f of the kth most visited location follows the following distribution $f(k) \sim k$ to the power of Q, where Q is 1.2 ± 0.1. For the example shown in Figure 8.2 Q was set to be 1.0. This means the one hundred visits are distributed against rank as 32, 16, 10, 8, 6, 5... For more on human mobility distributions see:

Chaoming Song, Tal Koren, Pu Wang and Albert-László Barabási, 'Modelling the Scaling Properties of Human Mobility', *Nature Physics*, Vol. 7, (2010) pp. 713-719.

HOME MAKING IS NOT JUST HOUSEWORK

In the following article Felski links the everyday to home. 'Like everyday life itself, home constitutes a base, a taken-for-granted grounding, which allows us to make forays into other worlds.'

Rita Felski, 'The Invention of Everyday Life', *New Formations*, Vol. 39, (1999) pp. 15-31.

THE SOCIOLOGIST JOHN GRADY

John Grady is a professor of Sociology and the William Isaac Cole Chair in Sociology Anthropology at Wheaten College, Massachusetts. He has a particular interest in the sociology of the visual.

John Grady, 'In Sight: Visualizing Relationships in Daily Life', unpublished paper presented at the *International Visual Sociology Association Conference*, University of British Columbia, Vancouver, (2011).

IF THE PATTERN OF MOBILITY OF AN INDIVIDUAL

A recent study shows that combining mobility data with context allows this type of long-term predictability. The authors below point out that this predictability would allow location-based advertising: 'Need a haircut? In 4 days, you will be within 100 metres of a salon that will have a $5 special at that time'.

Adam Sadilek and John Krumm, 'Far Out: Predicting Long-Term Human Mobility', *Twenty-Sixth AAAI Conference on Artificial Intelligence*, (2012).

Chapter 9 - Data to Life

A PLATFORM IS AN INTEGRATED OFFERING

Larry Keeley used this definition of a platform during a meeting with Matt Reed in 2011. It also featured in one of his presentations earlier that year.

Larry Keeley, presentation at IBF Conference '*13th Annual Corporate Venturing and Innovation Partnering*' (Newport Beach, 2011).

NON-COMMUNICABLE DISEASES

The following paper brings together many of the recent statistics on this and develops a number of mathematical models.

Abdesslam Boutayeb and Saber Boutayeb, 'The Burden of Non-Communicable Diseases in Developing Countries', *International Journal for Equity in Health*, Vol. 4, No. 2, (2005).

OBESITY, SMOKING, POOR DIET, INSUFFICIENT EXERCISE

This World Health Organization (WHO) report below sets out the statistics, evidence and experiences needed to launch a more forceful response to the growing threat posed by non-communicable diseases.

World Health Organization, *Global Status Report on Non-Communicable Diseases 2010* (Geneva: World Health Organization, 2011).

CHALLENGING TO INSTIL LASTING CHANGE
A good review of different health behaviour change approaches is given by Abraham and Sheeran.

Charles Abraham and Paschal Sheeran, *Understanding and Changing Health Behaviour: From Health Beliefs to Self-Regulation*, ed. by Paul Norman, Charles Abraham and Mark Conner, 2nd edn. (Hove: Psychology Press, 2000), pp. 3-24.

ESTABLISH A HEALTHY TOOTH-BRUSHING REGIME
The following paper provides an outline of an industrial approach to behaviour change, which ranges from a process to design and develop behaviour change interventions, to the development and use of technology to shape and measure behaviour.

Jean-Paul Claesson, Sue Bates, Kellie Sherlock, Feroud Seeparsand and Richard Wright, 'Designing Interventions to Improve Tooth Brushing', *International Dental Journal*, Vol. 58, Issue S5, (2008) pp. 307-320.

SMOKING CESSATION PROGRAMMES
In the UK, the National Institute for Clinical Excellence (NICE) provides guidance to healthcare professionals. Its information on smoking cessation incorporates both pharmacological and behavioural approaches.

NICE Public Health Guidance 10, *Smoking Cessation Services in Primary Care, Pharmacies, Local Authorities and Workplaces, Particularly for Manual Working Groups, Pregnant Women and Hard to Reach Communities*, PH010 (2008).

THE CARBON DISCLOSURE PROJECT
The Carbon Disclosure Project works with investors and companies to reduce the risks posed by climate change. Figure 3 of the report below analyses 256 companies.

Carbon Disclosure Project, *Carbon Reductions Create Positive ROI: Carbon Action Report 2012* (London, 2012).
<http://www.cdproject.net >

Chapter 10 - Coelition

THE MOST PROFOUND TECHNOLOGIES DISAPPEAR
Mark Weiser (1952-1999) was a computer scientist who became a chief scientist at Xerox PARC. He saw ubiquitous computing as the third

wave of computing: 'The first wave of computing, from 1940 to about 1980, was dominated by many people serving one computer. The second wave, still peaking, has one person and one computer in uneasy symbiosis, staring at each other across the desktop without really inhabiting each other's worlds. The third wave, just beginning, has many computers serving each person everywhere in the world. I call this last wave "ubiquitous computing" or "ubicomp".'

Mark Weiser, 'The Computer for the 21st Century', *Scientific American*, Special Issue on Communications, Computers, and Networks, (1991).

WIRELESSLY CONNECTED OVER LONGER DISTANCES
For example, the recently developed Weightless protocol describes a new wireless, medium-distance, machine-to-machine (M2M) communication protocol. This protocol is ideal for data transmissions that occur due to the function of a machine rather than a person. An example would be a smart meter that regularly transmitted information about average energy use or created *ad hoc* communications on peak energy usage.

<http://www.weightless.org/>

SERVICE DESIGN PRINCIPLES
Service design is an important new innovation discipline. It applies the thought processes and tools of product design to the dynamic relationship that exists between an organisation that is providing a service of some sort and its customers.

Marc Stickdorn and Jakob Schneider, *This is Service Design Thinking. Basics—Tools—Cases* (Amsterdam: BIS, 2012).

ENGINEERING DESIGN PRINCIPLES
The engineering design approach used for the Coelition standard explicitly needed to incorporate human elements as well as robust engineering. It was inspired by the OS X operating systems of Apple. What Apple achieved with OS X was a combination of high-grade operating system engineering, including elements from earlier UNIX-based systems, and a best-in-class graphical user interface known as Aqua. What the Apple user sees and feels is intuitive and human: what the machine sees is robust engineering.

STANDARDS DEVELOPED BY HEALTH ORGANISATIONS

Continua Health Alliance is an open non-profit organisation of healthcare and technology companies seeking to establish a system of connected and interoperable solutions that can improve the quality of personal healthcare. Dossia is a health data interoperability standard that aims to transform the US healthcare system by developing and making widely available a health record system that individual patients can control and manage themselves.

<http://www.continuaalliance.org/>
<http://www.dossia.org/>

COELITION IS AN OPEN STANDARD

Andrew Updegrove runs a comprehensive website describing the way that standards-setting organisations can be created.

<http://www.consortiuminfo.org/>
<http://www.iso.org/>

THE TERM 'OPEN' CAN MEAN DIFFERENT THINGS

A comprehensive discussion of both open source and open standards is provided by Daniel Särefjord in the report below. He makes the following point:

'Consider for example the practical examples Linux, which is an open source operating system, and Linear Tape-Open (LTO), which is an open standard specification for storage media developed by IBM, Hewlett-Packard, and Quantum. One solution has a rather ideological and altruistic history (Linux), the other was formed to achieve a de-facto standard with the objective of securing the commercial position of its founders (LTO). One is *gratis, available for modification* and *encourages redistribution* (Linux), the other *costs money, leaves no room for modification* and *prohibits redistribution* (LTO). Both initiatives are communicated as "open"'.

Daniel Särefjord, 'Open Platform Design – Towards a Theoretical Framework and a Practical Toolbox', unpublished Chalmers University of Technology report, (Göteborg, 2006).

Index

About the authors

MATT REED is a visiting professor at the University of Ulster and an Open Innovation director at Unilever Research and Development. He holds a 1st class honours degree in chemistry from Salford University, a PhD in quantitative microscopy from the University of Liverpool and a Chalmers Executive MBA in technology management. Matt has published more than 25 peer-reviewed scientific papers and filed a number of patents. In 1997 he co-authored a practical textbook on quantitative microscopy that continues to sell and be cited to this day. Matt is a keen gardener and lives in North-West England with his family.

JOSS LANGFORD is an independent management consultant and technical director for Activinsights Ltd, a developer of human motion measurement instruments. He holds a 1st class honours degree in cybernetics and control engineering from the University of Reading. Joss has had designs recognised by DBA and D&AD and filed numerous patents in a range of technical disciplines. He was a contributing author of *Growth Champions*, an innovation management book published in 2012. Joss also farms sheep in South-West England where he lives with his family.

CPSIA information can be obtained at www.ICGtesting.com
Printed in the USA
LVOW13s1844100314

376777LV00007B/953/P